ROUTES, ROADS AND LANDSCA

ROUTES, ROADS AND LANDSCAPES

Edited by

Mari Hvattum and Janike Kampevold Larsen
The Oslo School of Architecture and Design, Norway

Brita Brenna and Beate Elvebakk
University of Oslo, Norway

Routledge
Taylor & Francis Group
LONDON AND NEW YORK

First published 2011 by Ashgate Publishing

Published 2016 by Routledge
2 Park Square, Milton Park, Abingdon, Oxfordshire OX14 4RN
711 Third Avenue, New York, NY 10017, USA

First issued in paperback 2016

Routledge is an imprint of the Taylor & Francis Group, an informa business

British Library Cataloguing in Publication Data
Routes, roads and landscapes.
  1. Landscape assessment. 2. Geographical perception.
  3. Roads. 4. Roads in art. 5. Transportation.
  6. Transportation in art.
  I. Hvattum, Mari, 1966-
  304.2-dc22

Library of Congress Cataloging-in-Publication Data
Routes, roads and landscapes / by Mari Hvattum... [et. al.].
    p. cm.
  Includes bibliographical references and index.
  ISBN 978-1-4094-0820-8 (hardback)
  1. Landscape assessment. 2. Geographical perception. 3. Roads. 4. Scenic byways. 5. Automobile travel. I. Hvattum, Mari, 1966-
  GF90.R68 2011
  704.9'436--dc22
                                                                    2011012513

ISBN 13: 978-1-138-24614-0 (pbk)
ISBN 13: 978-1-4094-0820-8 (hbk)

Contents

SECTION II   THE ROUTE AS ICON AND OCCURRENCE

SECTION III   LANDSCAPES OF MOBILITY

CODA

List of Figures

Notes on Contributors

Gernot Böhme is Professor Emeritus in Philosophy at the Technische Universität Darmstadt, Germany, and Director of Institut für Praxis der Philosophie. His work covers topics like time, natural philosophy and phenomenology. In the last two decades he has focused particularly on ecological aesthetics, atmosphere and 'aisthetics'. His influential books include *Atmosphäre: Essays zur neuen Ästhetik*, *Architektur und Atmosphäre*, and *Aisthetik: Vorlesungen über Ästhetik als allgemeine Wahrnehmungslehre*.

Sarah Bonnemaison is an architect and Associate Professor at Dalhousie University. She holds a PhD in Human Geography, and has done extensive research in festival architecture and landscape history. Recent publications include *Installations by Architects* (with R. Eisenbach), *Architecture and Nature: Creating the American Landscape* (with C. Macy), and *Festival Architecture* (with C. Macy). She runs the architectural company FILUM together with Christine Macy, specializing in lightweight, collapsible structures and set design.

Brita Brenna is Associate Professor of Museology at the University of Oslo. She did her PhD on the Norwegian participation in nineteenth-century world fairs and international exhibitions. Her recent publications include *Technoscience. The Politics of Intervention* (co-edited with K.Asdal and I.Moser), and *Æmula Lauri. The Royal Norwegian Society of Sciences of Letters 1760–2010* (with H.W. Andersen, M. Njåstad and A. Wale). She is presently a senior researcher in the Routes project with a study of eighteenth-century topographical literature.

Tim Cresswell is Professor of Geography at Royal Holloway, University of London. His research interests centre on themes such as mobility and place/space, with a special focus on the ideological and political aspects of geographical concepts and relations. He is the author of *On the Move: Mobility in the Modern Western World* as well as *Place: A Short Introduction*, *The Tramp in America*, and *In Place/Out of Place: Geography, Ideology and Transgression*.

Vittoria Di Palma is Associate Professor at Columbia University, Department of Art History and Archaeology, where she works on eighteenth-century architecture and landscape, focusing particularly on connections between landscape and epistemology. She is the author of numerous articles on the subject and has recently published *Intimate Metropolis: Urban Subjects in the Modern City*, co-edited with D. Periton and M. Lathouri.

Beate Elvebakk has a PhD in Technology, Innovation, and Culture from the University of Oslo. She is Research School Coordinator for the Ethics Programme at University of Oslo, a senior researcher at the Norwegian Institute of Transport Economics, and a researcher in the Routes project. She has previously published in the fields of science and technology studies, and on the politics of road safety.

Lars Frers is a sociologist with his PhD from Technische Universität Darmstadt. A guest researcher to the Routes project, his work focuses on materiality and interaction. His recent publications include the book *Einhüllende Materialitäten: Eine Phänomenologie des Wahrnehmens und Handelns an Bahnhöfen und Fährterminals*, and the edited volumes *Encountering Urban Places – Visual and Material Performances in the City* (with L. Meier) and *Negotiating Urban Conflicts – Interaction, Space and Control* (with H. Berking, S. Frank, M. Löw, L. Meier, S. Steets and S. Stoetzer).

Torild Gjesvik is an art historian with her MA from the University of Bergen. She has taught art history, arts communication and curatorial studies at the University of Bergen and at Telemark College, and has practiced as a freelance art historian, critic and curator. Gjesvik is presently a PhD candidate in the Routes project, writing on the relationship between road design and artistic representation in nineteenth-century Norway at the Institute for Cultural Studies and Oriental Languages at the University of Oslo.

Mari Hvattum is Professor of Architectural History and Theory at The Oslo School of Architecture and Design. She holds a PhD in Architectural History from University of Cambridge, and an MA in Architecture from The Norwegian Institute of Technology. Her publications include the monograph *Gottfried Semper and the Problem of Historicism*, and edited volumes such as *Tracing Modernity: Manifestations of the Modern in Architecture and the City* (with C. Hermansen). Hvattum is project leader of the research project Routes, Roads and Landscapes: Aesthetic Practices *en route*, 1750–2015.

Charlotte Klonk is Professor of Art History at the Humbolt Universität zu Berlin. She holds a PhD from University of Cambridge and is the author of numerous publications on modern art, architecture and landscape such as *Science and the Perception of Nature: British Landscape Art in the Late Eighteenth and Early Nineteenth Centuries* and *Art History: A Critical Introduction* (co-authored with M. Hatt). Her recent book *Spaces of Experience: Art Gallery Interiors from 1800 to 2000* appeared in 2009.

Janike Kampevold Larsen holds an MA in comparative literature from University of Bergen and a PhD in comparative literature from the University of Oslo. Kampevold Larsen has worked extensively within the literary field in Norway as a critic, editor and consultant. Since 2008 she has been a post doc at The Oslo School of Architecture and Design, writing on the Norwegian Tourist Routes. Her recent publications include the monograph *Å være vann i vannet: Forestilling og virkelighet i Tor Ulvens forfatterskap*.

Christine Macy is Professor of Architecture and Dean of the Faculty of Architecture and Planning, Dalhousie University, Canada. Her published works include *Architecture and Nature: Creating the American Landscape* (with S. Bonnemaison), *Greening the City*, as well as the anthology *Festival Architecture* (with S. Bonnemaison). Her new book *Dams* was recently published by Norton.

Peter Merriman is Senior Lecturer in Human Geography at Aberystwyth University in Wales, UK. His research focuses on the themes of mobility and cultures of landscape, particularly the spaces of driving. He is a member of the editorial boards of the journals *Cultural Geographies* and *Mobilities*, and the author of *Driving Spaces: A Cultural-Historical Geography of England's M1 Motorway*.

David E. Nye is Professor of History at the Center for American Studies, University of Southern Denmark. He has edited and written numerous books including *Electrifying America*, *American Technological Sublime*, *Narratives and Spaces*, and *Consuming Power: A*

Social History of American Energies. Formerly President of the Danish Association for American Studies and Vice-President of the Nordic Association for American Studies, he co-edited the journal *American Studies* in Scandinavia from 1996 to 2003.

Finola O'Kane is an architectural historian and a historian of landscape, lecturing on architecture and conservation at University College Dublin. She is the author of *Landscape Design in Eighteenth-Century Ireland, Mixing Foreign Trees with the Natives*, and is currently completing her second book entitled *Ireland and the Picturesque*. Her research interests include Irish landscape history, architectural and landscape conservation and Irish social history.

Even Smith Wergeland is an art historian with an MA from the University of Bergen. He has taught historiography of modern architecture and visual rhetoric at the University of Bergen, and worked at Rogaland Museum of Fine Arts, Stavanger. His latest engagement was at the city planning department of Stavanger municipality, where he worked as advisor and planner. Now a Routes project PhD, his research deals with the motorway as a work of art in twentieth-century urbanism.

Charles W.J. Withers is Professor of Historical Geography, University of Edinburgh. He works in the fields of the history of geographical knowledge and science. Recent publications include *Geography, Science and National Identity: Scotland since 1520*, *Placing the Enlightenment: Thinking Geographically about the Age of Reason*, and the edited volume *Georgian Geographies: Essays on Space, Place and Landscape in the Eighteenth Century*. He is presently co-editing a volume on *Geographies of the Book*, and his *Geography and Science in Britain 1831–1939: A Study of the British Association for the Advancement of Science* appeared in 2010.

Thomas Zeller is Associate Professor at the University of Maryland, Department of History. Working within the fields history of technology and environmental history, Zeller has written numerous books, among others *Driving Germany: The Landscape of the German Autobahn 1930–1970*, and the co-edited volumes *How Green were the Nazis: Nature, Environment and Nation in the Third Reich, Germany's Nature: Cultural Landscape and Environmental History, The World behind the Windshield: Roads and Landscapes in the United States and Europe*, and *Rivers in History: Perspective on Waterways in Europe and North America*.

Acknowledgements

This collection is part of an on-going research project financed by the Norwegian Research Council: *Routes, Roads and Landscapes, Aesthetic Practices* en route, *1750–2015*. In the autumn of 2009, the Routes team gathered a group of scholars in Oslo for a two-day exchange on routes, roads and landscapes. The extraordinary material they brought with them and the generosity with which they shared their thoughts inspired us to make a book, inviting a few more of our favourite routes scholars along the way. To our astonishment, they all accepted. We wish to thank the authors for their insights, generosity and patience. We also wish to thank colleagues in Norway and abroad who have helped, supported, and inspired the Routes project in so many ways, particularly Karl Otto Ellefsen and Mari Lending at The Oslo School of Architecture and Design, Kristina Skåden and Liv Emma Thorsen at the University of Oslo, Marie-Theres Fojuth at the Humboldt Universität zu Berlin, Kenneth Olwig at The Swedish University of Agricultural Sciences, Bruno de Meulder and Kelly Shannon at KU Leuven, Frederik Tygstrup at The University of Copenhagen, Anne Katrine Geelmuyden at Norwegian University of Life Sciences, Thor Bjerke at the Norwegian Railway Museum, and Geir Paulsrud at the Norwegian Road Museum. Many thanks also to Ingrid Book and Carina Hedén for their inspiring road photography exhibition shown at AHO during the Routes conference. We are grateful to the Norwegian Research Council for their long standing support, to the Oslo School of Architecture and Design for hosting the Routes project, and to the Norwegian Institute of Transport Economics, the Institute for Cultural Studies and Oriental Languages at the University of Oslo, and the Norwegian Public Roads Administration for their participation and support. Lastly, many thanks to Ashgate and Valerie Rose for efficient, intelligent, and professional guidance throughout the publishing process.

Introduction: Routes, Roads and Landscapes

Mari Hvattum, Brita Brenna, Beate Elvebakk, Janike Kampevold Larsen

Culture is more about routes than roots.[1]

In his 1780s bestseller *Theorie der Gartenkunst*, C.C.L. Hirschfeld declared that '[t]he foremost purpose of the path is to conduct us to notable scenes with no need for retracing one's steps'. Unconcerned with its more mundane uses, Hirschfeld considered the path an aesthetic orchestration – a device for seeing and experiencing the landscape:

> [...] the paths allow not only for the enjoyment of variations and diverseness but also for the most advantageous disclosure of the best vistas, sometimes suddenly, sometimes gradually, while on the other hand concealing any unpleasant sights. Therefore, the laying out of paths demands the most careful attention to the point of view from which the object strikes the eye.[2]

Hirschfeld spoke about garden paths, yet his insistence on an intimate and reciprocal connection between the landscape and the moving subject has relevance even outside the domain of the garden. Considering routes key to the appreciation and comprehension of the landscape, he anticipated a persistent theme in modern cultural history, a theme that would resurface again and again, be it in nineteenth-century railway literature, twentieth-century road studies, or twenty-first-century theorizing of place, space and movement.

More notable for his typicality than his originality, Hirschfeld's musings on paths echoed a long tradition. From Joseph Addison to William Gilpin and from Alexander Pope to Claude-Henri Watelet, the walk had long been conceived as a perceptual participation in the landscape – an involvement both forming and transforming its surroundings.[3] The Frenchman Watelet put it perhaps most lucidly: 'Indeed, nothing is more like the progress of our thoughts than these paths men create in the spacious countryside [...]. The person viewing picturesque scenes in a park [...] changes their organization by changing his location'.[4] Acutely sensitive to the experience *en route*, eighteenth-century garden theory anticipated the 'view from the road' so vividly described by twentieth-century authors such as Appleyard, Lynch and Myer, J.B. Jackson, Alison Smithson, Wolfgang Schivelbusch, and many more.[5] Described not only in words, of course:

Figure I.1 C.C.L. Hirschfeld, Theorie der Gartenkunst *(1779–85)* (Leipzig: Weidmanns, 1779–85), vol. 2, p. 160.

the relationship between routes and landscapes had been elaborated for centuries in garden design and agriculture, road building and city planning, landscape painting and architecture – making routes and roads an apt prism for studying the ways in which the modern landscape has been made, seen, used, and understood.

This collection examines routes and landscapes and the many ways in which they are entangled. It looks at how movement has been facilitated, imagined and represented, and how such movement in turn has conditioned the modern western understanding of the landscape. If 'landscape' is defined as culturally configured nature, then infrastructure may be considered the single most important factor generating it.[6] Routes make nature accessible, defining our viewpoint towards it and conditioning our involvement with it. Paths, roads, canals and railway lines constitute poignant linkages between nature and culture, representing as well as ordering our relation to the natural world. Georg Simmel spoke about 'the miracle of the road' and considered road building a defining human characteristic, laden with cultural meaning.[7] Despite their significance, however, infrastructural 'objects' such as roads have received surprisingly little

Figure I.2 Road building in Northern Norway during WWII (photographer unknown)
Reproduced with permission of the National Library of Norway, archive folder Norwegen 1945/NB SKM 04 20 0063a.

scholarly attention beyond their historical or utilitarian aspects. This collection is part of a small but expanding literature aiming to correct this oversight.[8] Suggesting that our sense of the landscape is profoundly conditioned by the way we move through it and across it, the chapters that follow bring to attention the rich cultural, historical and aesthetic significance of routes, roads, and landscapes. They investigate the ways in which various kinds of routes have shaped modern conceptions and representations of the landscape, and inquire into the role of the route itself, both as a material object and as a setting for a wide range of material, cultural, and aesthetic practices.

To be sure, the interest in movement through the landscape is part of a wave of scholarship that has emerged in the last decades, reassessing the relationship between movement and matter and criticizing the 'metaphysics of fixity' so pervasive – or so we are told – in the modern western world.[9] Social scientists have proclaimed a 'mobilities turn' and identified a 'new mobilities paradigm' as a way to grasp a fast moving present.[10] Historians and cultural historians speak about networks and relations rather than facts, and art historians have long abandoned their traditional object centred approaches and turned their attention to artistic processes. Artists, architects and landscape architects, similarly, focus increasingly on dynamic design processes and parametric systems rather than on 'works' in a traditional sense. This collection could perhaps be considered a part of such a 'mobilities turn', yet the contributions that follow are less concerned with paradigm shifts than with meticulous, micro-historical inquiry, tracing specific routes through particular landscapes at given times, and discussing their relationships and significances. Doing that, the collection poses a tacit challenge to the 'new' mobility paradigm's claim to newness, pointing instead to the intricate ways in which movement and mobility have informed people's perceptions of the world around them for centuries. Not only has movement shaped our understanding of the landscape, it has shaped the actual landscape as well, fundamentally structuring the world around us as a political, social and physical territory with a long history and an even longer *Wirkungsgechichte*. It is these kinds of processes – historical, epistemological and physical – that are under scrutiny here.

Such a heterogeneous material requires a multifaceted approach, and the essays in this collection represent a number of disciplines and outlooks. The sociologist John Urry has suggested that the mobility turn is 'post-disciplinary', meaning presumably that traditional disciplinary boundaries dissolve as one tries to grasp such a ubiquitous phenomenon as movement. Our strategy is different. Rather than dissolving disciplinary specificity we cultivate it, aiming to study routes and roads from as many angles as possible. We believe that each discipline – and this collection encompasses at least half a dozen – gleans different things from its material, and that together they may fill in each other's blind spots. It means that we align approaches not normally found side by side (such as Section III's heady mix of cultural geography, aesthetics and road planning), but we do not aim to fuse or obliterate them. The collection remains in that sense doggedly and unfashionably multi-disciplinary (rather than the more trendy post-, trans-, or inter-), honouring the vocabulary and methodology of each contribution. Having said that, the topic does invite transgression. Routes and roads tend to wind their way through the most surprising material, inevitably entailing meaning *and* materiality, sensation *and* discourse, history *and* contemporary practice. It forces the art historian to engage with planning procedures, the literary theorist to study road cuts and drywall techniques, and the social scientist to tackle aesthetic theory. It brings together dimensions that are all too often separated in contemporary academic discourse, reminding us of the concrete materiality that grounds even the most abstract phenomenon and the historical depth that underlies even the most ephemeral contemporary practice. Drawing from cultural history, sociology, landscape history, geography, art history, science and technology studies, architectural history, literary theory and philosophy, the chapters challenge each other's remits while sharing generously their outlook and approach.

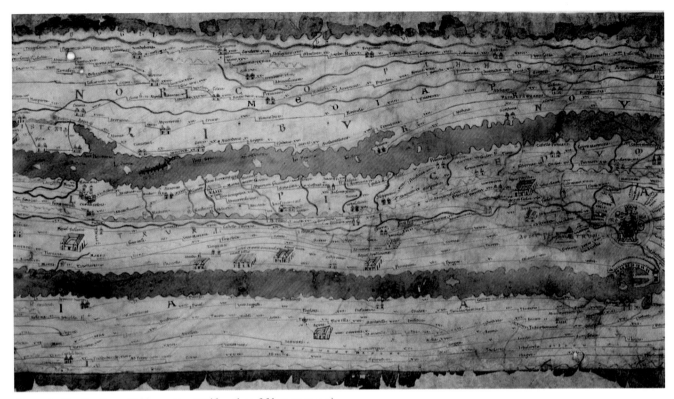

Figure I.3 The Peutinger Table, section IV (fourth to fifth century AD)
Reproduced with permission of the Österreichische Nationalbibliothek, ÖNB/Vienna.

Anyone who reads Pausanias' *Description of Greece* (second century AD) or looks at the Roman Peutinger Table – the earliest known route map – will know that the entanglement of routes and landscapes is nothing new.[11] The Renaissance was full of movement through landscapes (remember *Hypnerotomachia Poliphili*, or the gardens of the Villa d'Este), and the Baroque made movement a first principle of design, both of cities and landscapes.[12] Maps have facilitated movement through the landscape for at least four thousand years, and have at the same time moved the landscape itself, making it transportable through various inscription techniques. When this collection sets off in the mid eighteenth century, however, it is because enlightenment cartography and notions of 'improvement' rendered the

relationship between the route and the landscape particularly acute. As cultural historian Brita Brenna shows in her opening chapter, the landscape of the eighteenth century is gradually being seen, described, and depicted in new ways – ways that are intimately linked to new ways of knowing. Looking at journeys undertaken by the Danish king to Norway during the eighteenth century, Brenna charts this change. While the landscape descriptions from the early journeys focus almost exclusively on the king himself and his actual or symbolic movement through the territory, the later accounts are marked by a dual concern for the landscape as, on the one hand, an economic resource and on the other, a source of aesthetic enjoyment. Following in the wake of the king, Brenna's study records the emergence of a new way of knowing the territory, one that oscillates between usefulness and beauty – pragmatism and aesthetic sensibility. This modern dialectics is pursued by several authors in Section I; 'Moving through the landscape'. In her careful analysis of eighteenth-century river routes, architectural historian Vittoria Di Palma's shows how the river in eighteenth-century Britain became the harbinger of a fluvial aesthetics – 'an aesthetics of a mobile eye on a winding route'. Using William Gilpin's river studies and John Ogilby's strip maps as her main material, she demonstrates how movement came to play a key role in eighteenth-century landscape experience and representation.

Di Palma's interest in routes in relation to eighteenth-century aesthetic theory is complemented by landscape historian Finola O'Kane who looks at road construction in eighteenth-century Ireland. O'Kane shows how roads played a key role in the formalization of the eighteenth-century landscape. Serving aesthetic, economic and colonial agendas, the road was a means to create an improved, well-tempered landscape that corresponded to the tourist's expectations. In an orchestrated attempt to accommodate the foreign view on Ireland, roads and routes both formed and transformed the national landscape.

The intimate connection between routes, landscapes and landscape painting is pursued further in art historian Torild Gjesvik's contribution, where we enter the dining room of Peder Anker – Intendant General of roads in Norway – around 1800. Looking carefully at the landscape paintings on the wall, Gjesvik shows how the late eighteenth-century road landscape brings together the traveller, the view, and the road builder, thus facilitating a complex interweaving of social, political and aesthetic concerns. The road, here, is an aesthetic orchestration and a political tool, but also a meeting place and arena for different social practices.

The appreciation of landscapes underwent intricate but profound transformations around the turn of the eighteenth century. Art historian Charlotte Klonk chooses an unexpected source for investigating these changes: plant physiognomy. Klonk traces the close connection between plant physiognomy and landscape representation around 1800, highlighting how different aesthetic presumptions allowed for different understandings of plants – and landscapes. Looking specifically at Alexander von Humboldt's journey to South America, she shows how the physiognomy of plants served as a clue to understanding, not only the individual botanical specimen, but also the new territory encountered *en route*. Cultural geographer Charles Withers explores further the matter of scientific expeditions and the particular kind of knowledge they involve. Following the British seafarer Captain George Francis Lyon, Withers examines the relationship between field work, field notes, and published accounts from Lyon's African, American and Arctic journeys from 1819 to 1832. The 'route to truth' becomes a fundamental question here: how may the inevitable discrepancies between experience and representation be understood in these accounts? By examining the multiple materialities of narrating, Withers allow us to understand more fully the epistemic basis of *en route* writing.

Moving from the king's landscape of the early eighteenth century to the scientific explorers' landscape of the early nineteenth, Section I maps a profound change in Western attitudes to the landscape and the movement through it. Nowhere does this new attitude come more forcefully to expression than with the new modes of transport emerging in the nineteenth and early twentieth centuries. The train and the car made the landscape available in entirely new ways, opening it for practical and aesthetic exploitation on an unprecedented scale. Section II; 'The route as icon and occurrence', studies the emergence of this modern transportation landscape. In his opening essay, historian David Nye presents three key moments in this development: the early nineteenth-century canal networks, the railway landscape of the late nineteenth century and the contemporary road landscape. Seeing the different routes as expressions of competing aesthetic practices, Nye examines the way experiences of mobility were imbued with particular aesthetic significance. Focusing on the sense of travelling and the traveller's experience of sites travelled through and to, Nye analyses the modern transportation landscape and its perceptual impact.

The subsequent chapters in Section II take up individual strands in Nye's historical overview. Architectural historian Mari Hvattum pursues the particular view on the landscape afforded by the railway in the mid nineteenth century, using the landscape photographs of the Norwegian railway director C.A. Pihl as her material. Merging the dream of 'naturalness' with an ambition for technological progress, Pihl's photographs present an ideal landscape in which nature and technology were brought together into a new totality. Similar issues are at stake in historian Thomas Zeller's study of 1930s parkways in Germany and the United States. These roads and roadside sceneries were created specifically to be seen from a moving car, constituting a carefully constructed 'world beyond the windshield' in which the tension between the given and the man-made was attempted resolved.[13] Comparing Blue Ridge Parkway in Virginia/South Carolina with the Deutsche Alpenstraße extending 450 kilometres on the northern crest of the Alps, Zeller's meticulous study demonstrates the crucial role these roads played in the construction and orchestration of national identity. The architects and architectural historians Sarah Bonnemaison and Christine Macy query further into American road history by looking at Benton MacKaye's interwar theories of urbanization. His idea of the townless highway, implemented for the first time in the Tennessee Valley Authority project, drew on parkway ideals as well as theories of landscape preservation developed during MacKaye's work on the Appalachian trail. MacKaye envisioned the highway as a planning tool and a vital part of the 'flow control' that was to regulate infrastructure, movement, and population patterns in the Tennessee Valley. With its comprehensive system of dams, highways, landscapes and settlements, the TVA project represents an ambitious attempt at reconciling the experience of wilderness with progressive, and highly controlled, social ideals.

The chapters in Section II study the cultural significance of actual routes, showing infrastructure as a complex entanglement of meaning and matter. Twentieth-century automobility also harbours a strong utopian element, however, explored by art historian Even Smith Wergeland in the last chapter of Section II. Smith Wergeland looks at nineteenth and twentieth-century 'autopias', drawing a line from Albert Robida's utopian images of mobility in the novel *The Twentieth Century* (1890), through Norman Bel Geddes *Futurama*-exhibition for General Motors in 1939, to the Dutch architectural practice MVRDV's *Skycar City* project from 2007. Looking at the way this utopian imagery has impacted architecture and urbanism in the twentieth century, Smith Wergeland points to the surprisingly static thinking underlying modern utopias of mobility.

Drawing from rich and diverse historical materials, the chapters in Section II study routes and visions that have shaped the modern landscape, simultaneously constituting it *qua* landscape and making it accessible for practical and aesthetic exploitation, reification and interaction. Underlying these occurrences are processes of movement and mobility that constitute significant objects of study in their own right. The third section; 'Landscapes of mobility' focus on such processes, looking specifically at the cultural and aesthetic significances invested in mobility in the twentieth and twenty-first centuries. The cultural geographer Tim Cresswell sets out by scrutinizing and expanding the 'new mobilities paradigm', defining mobility as an entanglement of movement, representation and practice. He argues for a politics of mobility capable of thinking beyond 'mobility' versus 'immobility', encompassing speed, motive force, rhythm, route, experience and friction. Contemporary mobilities studies should pay heed to historical forms of mobility, Cresswell argues, as well as to forms of fixity, stasis and immobility. The interchange between movement and materiality *en route* is explored further by literary scholar Janike Kampevold Larsen in her critical analysis of viewpoints and rest stops along Norwegian tourist routes. Investigating the way these installations exhibit, interpret, and mirror nature, Larsen regards the Norwegian Tourist Route Project as a curatorial practice patterned largely on a nineteenth-century aesthetic, in which potent visual experience is privileged above more complex, sensual interactions with the landscape. Sociologist Lars Frers continues on the topic of sensuous experience along the road, albeit with quite a different perspective. Charting the tourist's bodily experiences, Frers explores the physical processes taking place at view points and rest stops. He shows how the tourists' experience of the landscape is in a constant and intimate interplay with more mundane urges and need – functions that usually fall well below the radar of academic enquiry. Philosopher and science and technology scholar Beate Elvebakk also explores a topic rarely discussed in academic scholarship, giving us a glimpse into the theory and practice of contemporary road planning. Looking particularly at the aesthetic guidelines of the Norwegian Public Roads Administration, Elvebakk asks what is considered a beautiful road today, and how it is actually planned. Through a close reading of planning documents, Elvebakk unravels the aesthetic presuppositions informing contemporary road planning, demonstrating, among other things, how the eighteenth-century insistence on variety lives on in present day's bureaucratic prescriptions of one visual diversion for every three minutes of driving. Hirschfeldt's pronouncement that roads and paths 'demand the most careful attention to the point of view from which the object strikes the eye' still holds, it seems, in twenty-first-century road design.

Tim Cresswell called for mobilities studies capable of tackling the cultural complexity of mobility and movement. In his detailed study of the English M1 motorway, human geographer Peter Merriman provides a good example of how such studies may proceed. Merriman argues that motorways cannot be understood either as 'non-places' or as linear structures. Using the metaphors 'gathering' and 'enfolding' he presents the motorway as topological and relational spaces that create new relations and geographies, constantly

Figure I.4 Ingrid Book and Carina Hedén, road cut along E6, 2009
Reproduced with permission of the artists.

forming and emerging through the flow of bodies, vehicles and materials.

The relationship between roads and landscapes has all too often been understood in purely visual terms, as if the 'panoramic perception' of the modern traveller is the only way to access nature *en route*.[14] If this heterogeneous collection has a common aim, it must be to show that routes are not only a visual orchestration of views and vistas but also means to participate in the landscape, both epistemologically and perceptually. Joseph Addison expressed this beautifully in the opera *Rosamond* (1707), where his heroine sings passionately about walking through the forest:

From walk to walk, from shade to shade,
From stream to purling stream convey'd,
Through all the mazes of the grove,
Through all the mingling tracts I rove,
Turning,
Burning,
Changing,
Ranging,
Full of grief and full of love.[15]

With Addison's sumptuous ramble in mind, it is fitting to have the philosopher Gernot Böhme conclude the collection by pointing to the journey and the route not only as an aesthetic but *aisthetic* interchange between movement and matter. Böhme outlines a notion of the modern landscape, not as a realm of alienation and disembodied experience, but rather as a historically, culturally and sensuously charged territory. Doing that, he brings the discussion back to the Routes collection's main concern: the movement through the landscape and the way it affects our perception, construction, use and understanding of the world around us.

ENDNOTES

1 Paraphrasing Tim Cresswell's more cautious assertion: 'Culture, we are told, no longer sits in places, but is hybrid, dynamic – more about routes than roots'. *On the Move. Mobility in the Modern Western World* (London: Routledge, 1996), p. 1.

2 Christian Cay Lorenz Hirschfeld, *Theorie der Gartenkunst*, 5 vols (Leipzig: Weidmanns, 1779–1785), abbreviated and translated by Linda B. Parshall, *The Theory of Garden Art* (Philadelphia: University of Pennsylvania Press, 2001), p. 251. The editors are grateful to garden historian Madeleine von Essen for bringing Hirschfeld to our attention.

3 See for instance Joseph Addison's essay 'The Pleasures of the Imagination', *The Spectator* (25 June 7012), Alexander Pope's *Moral Essays* to Lord Burlington at Chiswick (London, 1731) and Lord Cobham at Stowe (London, 1734) – two of the most celebrated landscape gardens in early eighteenth-century England. See also William Gilpin's *Essay on Picturesque Travel* (London: Blamire, 1792), and Claude-Henri Watelet's *Essai sur les jardins* (Paris: Prault, 1774). Rebecca Solnit touches on aspects of this tradition in her *Wanderlust. A History of Walking* (London: Penguin, 2001), as does Tim Richardson, *The*

Arcadian Friends: Inventing the English Landscape Garden (London: Bantam, 2007).

4 Claude-Henri Watelet, *Essay on Gardens*, trans. and ed. Samuel Danon (Philadelphia: University of Pennsylvania Press, 2003), p. 26 and p. 37.

5 Donald Appleyard, Kevin Lynch and John R. Myer, *The View from the Road* (Cambridge, Massachusetts: MIT Press, 1964). John Brinckerhoff Jackson's 1950s essays on roads and landscapes are collected in e.g. Helen L. Horowitz (ed.), *Landscape in Sight. Looking at America* (New Haven: Yale University Press, 1997). Alison Smithson, *AS in DS: an Eye on the Road* (Baden: Lars Müller, 2001), Wolfgang Schivelbusch, *The Railway Journey: The Industrialization of Time and Space in the 19th Century* (Berkeley: University of California Press, 1986).

6 See e.g. John Dixon Hunt, *Greater Perfection: The Practice of Garden Theory* (London: Thames & Hudson, 2000). Kenneth Olwig accounts for the rich history of the concepts of 'nature' and 'landscape' respectively in *Landscape, Nature and the Body Politics* (Madison: University of Wisconsin Press, 2002).

7 Georg Simmel, 'Bridge and Door' (1909) in David Frisby and Mike Featherstone (eds), *Simmel on Culture* (London: Sage, 1997), p. 171.

8 The interest in the cultural significance of infrastructure is growing internationally, as indicated by e.g. Marc Desportes, *Paysages en mouvement. Transports et perception de l'espace, XVIIIᵉ–XXᵉ siècle* (Paris: Gallimard, 2005), Christine Macy and Sarah Bonnemaison, *Architecture and Nature: Creating the American Landscape* (London: Routledge, 2003), Thomas Zeller, *Driving Germany: The landscape of the German Autobahn 1930–1970* (Oxford: Berghahn, 2007), Christof Mauch and Thomas Zeller (eds), *The World Beyond the Windshield. Roads and Landscapes in the United States and Europe* (Athens: Ohio University Press, 2008), Peter Merriman, *Driving Spaces: A Cultural-Historical Geography of England's M1 Motorway* (London: Blackwell, 2007),

Joe Moran, *On Roads, a Hidden History* (London: Profile Books, 2009), and Cresswell, *On the Move*. The relationship between architecture and mobility is another emerging topic in contemporary scholarship, indicated by for instance Mitchell Schwartzer, *Zoomscape* (New York: Princeton Architectural Press, 2004), Sarah Bonnemaison and Christine Macy, *Festival Architecture* (London: Routledge, 2008), and Jilly Traganou et al. (eds), *Travel, Space, Architecture* (Aldershot: Ashgate, 2009).

9   Cresswell, *On the Move,* ch. 2.

10   John Urry, *Mobilities* (Cambridge: Polity, 2007), chs 1 and 2.

11   Pausanias, *Description of Greece* (second century AD), trans. W.H.S. Jones (London: Heinemann, 1965–71). *Tabulae Peutingeriana* is a medieval copy of a Roman map dating from late fourth or early fifth century AD. It encompasses 11 sheets of parchment, 33 cm high and with a total length of 680 cm. The map shows towns, seas, and topographical features such as forests and mountain ranges, but its most striking feature is the 200,000 kilometres of roads, crisscrossing the Roman world from Iceland to Sri Lanka. The editors are grateful to Vittoria Di Palma for bringing the Peutinger Table to our attention.

12   Francesco Colonna, *Hypnerotomachia Poliphili: The Strife of Love in a Dream* (1499), trans. Joscelyn Goodwin (London: Thames & Hudson, 1999). For an original study of the gardens of the Villa d'Este, see David Dernie and Alastair Carew-Cox, *The Villa d'Este at Tivoli* (London: Academy Editions, 1996). An interesting example of the study of early routes is Hans Bjur and Barbro Santillo Frizell, *Via Tiburtina – Space, Movement and Artifacts in the Urban Landscape* (Rome and Stockholm: The Swedish Institute in Rome, 2009). Sarah Bonnemaison and Christine Macy's *Festival Architecture* presents incisive studies of routes and movement in Renaissance and Baroque architecture.

13   *The World beyond the Windshield* is the title of Mauch and Zeller's collection of essays on roads and landscapes, which has greatly inspired this anthology.

14   For a critique of modern pictorialization of the landscape, see Gina Crandell, *Nature Pictorialized, 'The View' in Landscape History* (Baltimore and London: Johns Hopkins University Press, 1993).

15   *Rosamond* was set in Woodstock Park in Oxfordshire, which in 1707 formed part of Vanbrugh's emerging Blenheim Palace and park. Joseph Addison, *Rosamond. An Opera* (London: Tenson, 1707), Act I, Scene 2, p. 8.

Section I   Moving through the Landscape

This section discusses pictorial and textual representations of routes and landscapes, ranging from eighteenth-century topographical literature and landscape gardens, to the multifaceted landscapes of early nineteenth-century expeditions. Oscillating between scientific investigations, colonial conquest and romantic reverence, these descriptions and depictions make up a vivid testimony to the reciprocal relationship between the landscape and the route.

1 King of the Road: Describing Norwegian Landscapes in the Eighteenth Century

Brita Brenna

In the 1920-publication *The Artistic Discovery of Norway* we meet Norwegian landscapes, drawn, painted and etched by the artist Erik Pauelsen after his tour through Norway in 1788. The author of this publication, the renowned cultural historian Carl W. Schnitler, hails Pauelsen as the first to give an artistic depiction of Norwegian nature. 'In Erik Pauelsen's drawings we trace for the first time an agitated mind and a willingness to express the greatness and beauty of what he sees – in other words a feeling for nature'.[1] From the inception of the nineteenth century, Norwegian landscape painting was instrumental in producing a national imagery and instilling a sense of a unique Norwegian nature, connected to the creation of a Norwegian state in 1814 after 400 years of Danish rule. By the beginning of the twentieth century, Norwegian nature and landscape imagery had been naturalized to the degree that one could talk about the artistic *discovery* of the Norwegian landscape. To query in what sense this could be termed a 'discovery', the path followed in this article heads towards what existed before this discovery.

On a drawing made by a local clergyman in a parish in the eastern part of Norway in 1685, we meet the king of the road (Figure 1.1): The absolutist monarch king Christian V towers over the valley landscape with snow-clad mountain-tops, pleasant pine trees and a sun emanating rays over this king of God's grace. In 1685 the absolutist king made his seven weeks long royal progress through Norway, the Northern part of the state of Denmark-Norway. The local clergyman Henning Munch used the rare occasion to present the king with a so-called 'supplikk', a letter of humble request, where he asked for funds to keep his parish church in good repair. On the top of the 'supplikk' he had made a sketch of the king overlooking the valley and taking possession over the territory.[2] The king has the presence of an equestrian statue, similar to the actual statue produced in Copenhagen at the same time, to be placed on the central square of the

Figure 1.1 This drawing is part of a letter to the king, a 'supplikk', drawn and written by the clergyman Henning Munch in 1685 to ask for means to mend his parish church (photo by Leif Stavdahl) Maihaugen, Lillehammer: SS-DS-1995-0022. Reproduced with permission from Maihaugen.

town, Konges Nytorv. In Munch's sketch, however, the king is not in command of the state, the capital Copenhagen, demonstrating his power over his subjects, but of nature, in the form of a Norwegian landscape.

Munch's drawing frames this article, which will discuss landscape as a product of relations of power and knowledge in early modern Norway. A network of different practices and actors took part in establishing a Norwegian territory politically, scientifically and aesthetically. The material is scattered and varied, and this heterogeneity is of importance to the argument, which holds that different practices were important for what would later be named 'a feeling for nature' and an 'artistic discovery of Norway'. The visual landscape descriptions produced in the wake of the kings' journeys through the country during the eighteenth century is the most important material, together with the landscapes written into topographic literature from the middle of the eighteenth century onwards.

In his article 'Knowledge of the Territory', historian Jacques Revel investigates maps and proto-statistics as ways of knowing the territory in early modern France. Spanning a long period from the thirteenth to the nineteenth century, the article surveys efforts by public authorities to describe and homogenize a national territory, only to find that these were heterogeneous and contradictory for a long period of time.[3] This argument is relevant also for the material I am researching, even if I am more interested in the visual character of this territorial knowledge. I will look at how visual descriptions of landscape were connected to different kinds of political, scientific and aesthetic practices for charting territory, or rather how these practices were entangled. One large shift is evident, namely the transition from a description of territory for the king to a description of territory for the citizens. But within this larger transformation there were strategies with quite divergent genealogies and often contradictory results.

The caption on the above drawing reads *Invia gratia nulla est via*. This is a word-play on Ovid's *Metamorphoses*, book 14, where Anaeas asks for help to get through to the underworld to meet the ghost of his father, and is answered by the sentence *Invia virtuti nulla via est*, 'no way to noble vertue is denide', in the words of a seventeenth-century English translation.[4] In Munch's drawing, grace is substituted for virtue, and the translation can be something like: 'No way is impassable for grace'. We could translate it even more freely, as 'All roads are accessible for the king of God's grace'. The king in Copenhagen ruled over a conglomerate state, dispersed territories, with the capital Copenhagen as the uncontested political, religious, economical, and cultural centre. Norway was ruled by the king in Copenhagen, but had approximately the same number of inhabitants, a much larger territory and a very different topography. As opposed to the flat and fertile fields of Denmark, Norwegian territory consisted of forests and mountains. Norway possessed rich natural resources – fish abounded in the North Sea, large amounts of timber were shipped from ports along the coast, and in the mountains there were metals and minerals. These were sources of income for the king, but the country was undeniably a vast, unknown and uninviting territory. If Munch's caption alluded to the context of the proverb in Ovid, the message was that travelling through Norway was like descending into an underworld inhabited by the ghosts of heroes.

THE KING'S PROGRESS

The territory was accessible for the king of God's grace, but travel was filled with hardships, and removed the king from the centre of the kingdom for months. Kings would regularly visit the different parts of the kingdom, but their visits to Norway were not exactly frequent.[5] During the whole eighteenth century, sovereigns toured the country only four times, in 1704, 1733, 1749 and 1788. The 1749 tour was limited to the eastern region.[6]

During one full century, most of what was produced of landscape depictions from Norway came about through the different kings' journeys through the country.[7] Norwegian painters and draftsmen were a scarce commodity, and most of the ornamental work produced in churches and wealthy homes as well as portraits produced during this period were executed by travelling artists of Swedish, German and Dutch origin.[8] Landscapes were not in demand, and no significant activity in mapping or prospect making can be documented. Thus, the largest collection of landscapes of early modern Norway came about as a result of the trip made by King Christian VI and Queen Sophie Magdalene in 1733. They brought with them approximately 190 people, and it has been said that rather than travelling through the country as a regent, the king was moving his court 'over mountains, rivers and inlets'.[9] The queen's presence was unprecedented and made for the size of the company. They travelled from Copenhagen through Jutland up to Flatstrand (now Frederikshavn) before boarding nine ships that took them to Christiania, where the governor resided.[10] After visiting towns around the fjord, they travelled inland through Hamar, over Dovre to Trondheim. Part of the company took the same route back to Christiania, while the king and the rest of the company went by small boats down the coast and back again to Larvik. After 4 months and 11 days, the court was back in the Royal palace.

MAP OF THE ROAD

Three maps and one lavishly illustrated manuscript remain from this journey, where the route of the royals forms the main topic.[11] The maps were most certainly made after the tour, to commemorate the journey. These are impressive both in figurative content and in size, the largest measuring 142 x 520 cm, and beautifully hand-coloured. Roads, rivers, bridges and the most strenuous passages are depicted, the last ones by 'perspective', as indicated in the title of the maps.[12] On the largest map, seven of the difficult passages are drawn, as well as seven plans of cities and fortresses. There is also a 'record' giving the details of the day to day movement of the entourage from 2 June to 28

Figure 1.2 This large scale map measures 142 x 520 cm and gives a detailed account of the King's route from 2 June to 28 July 1733

The Royal Library, Copenhagen: KBK 1112-0-1733/3. Reproduced with permission from The Royal Library, Copenhagen, Department of Maps, Prints and Photographs.

July in the first two maps, between 28 July and 25 August on the third. One map is heavily worn, but the second is a copy in good condition, with a cartouche which differs from the original and with a text giving the names of Sergeant Torban Knorr and Corporal Steffen Nicolay Holstein. This also states that the first map was made in 1733, the second 'copied and represented' in 1735. The third map is smaller (161 x 275 cm) and is basically representing the Norwegian coastline, but with the East upwards, so that the journey of the king is represented horizontally from West to East. The main focus here is on the difficult passage Manseidet, where the entourage had to leave the boats to pass over a mountain. Thus, the king's journey can be followed day by day in the texts on these maps, looking at the map we locate the exact position from day to day, but the main focus in the perspective drawings is on how the king could be seen moving through the landscape.

Taking the largest of the map as an example (Figure 1.2) we see how different systems of representations are combined: the fortresses and cities are depicted as plans, the difficult sequences of the road are represented in perspective, and the map itself in birds-eye view. The format leads the viewer's gaze along the route from bottom to top, indicating a progressive movement. The mountains look terrifying, the roads dangerous, and trees are scattered about in the landscape.

How were such maps used? They were evidently not travel maps to bring along, and the size indicates that they could be hung in a large room. We do not know who got to see them, but maps did have an important role in many European courts, the fad for extensive maps as decoration having spread from the palaces of Italian princes to courts around Europe in the sixteenth and seventeenth century.[13] Maps were paraphernalia of court life, vehicles for the workings of the state and court. These maps had details of Norwegian topography, and fortresses of strategic importance were inscribed. As such they could be used for political and military purposes. The most striking features in these maps, however, are the depictions of the mountains and the mountain passes over and through which the king traversed. Given this focus, the maps were presumably of little strategic interest. One could infer that knowledge of the territory was less important than knowledge about where the king had been, and that the presence of the king's body was more important for dominating the land than an abstract representation of the land itself. It was the king on Norwegian roads, the king's progress, that was to be commemorated in this ritual inscription of power.

MEMORY, HISTORY AND FAME

In the commemorative manuscript from Christian VI's journey in 1733, the first picture the king would have laid his eyes on is an allegory with mythical figures (Figure 1.3). Fama, the goddess of rumour and news, withdraws the crimson curtain, making 'the mildness and vigilance of the king towards Norway known to the whole world', according to the accompanying text. Memory holds up a mirror wherein the landscape that Fama uncovers is seen by History, who glances at the mirror as she engraves a marble pyramid with the words 'The Norwegian Journey'. This manuscript is a gift to the king, but Fama here reveals the news for memory and history, thus making the landscape of Norway known 'to the whole world'. This can be interpreted to say that the commemorative album was made for a wider public, even if it was a unique exemplar given to the king himself. Alternatively it is the inscriptions of History that will be spread. Interestingly, History seems to have only an indirect relation to the landscape, through the mirror of Memory. History is not made by experience, but by the act of commemoration.

The landscape uncovered by Fame, mirrored by Memory and represented by History has two basic features: rocky mountains and spruce trees. This, in short, is the Norwegian territory: spruce and rock. These are also the elements we saw in the drawing by the clergyman Henning Munch, but

whereas he pictured a cultivated or inhabited land with an unassuming road flanked by a fence, the frontispiece from the 1733 journey manuscript has no sign of human inhabitants. In the theatre of commemoration that this picture establishes, there is only an arid land without people to be remembered. The landscape is devoid of people, but the emblem of the king and queen are suspended on the sky, accordingly what is to be remembered and passed down as history is a landscape that has accommodated the royal couple.

The commemorative album consisted of 52 large folios, an introduction, a frontispiece, a detailed day-to-day description of the king's journey and a long closing poem. The poem is signed by Henrich Willemsen, who describes himself as a Norwegian and part of the king's company.[14] The text is accompanied by 68 non-paginated illustrations: maps, landscapes and gates of honour erected for the king. The gates of honour are the expressions of the ritual importance of the king's progress which had obvious parallels in other European countries. European kings had moved their courts to uphold control over territories and populations. On this Norwegian tour, the court was at work, receiving diplomats, performing the political tasks throughout the tour. But as elsewhere in Europe, there were also important rituals connected to the progress itself.[15] Different ranks played different parts in the confirmation of the king as their sovereign. When arriving in Christiania (now Oslo), the king and queen passed through a gate of honour with extensive allegorical figures as well as short proverbs, stating for example that the gate is open for the

Figure 1.3 Frontispiece of the manuscript celebrating King Christian VIs 1733 journey through Norway (photo by Arthur Sand)
We see Fama withdrawing the crimson curtain, Memory mirroring the landscape, and History engraving the monument while looking at the representation in Fama's mirror. *Norske Reise Anno 1733. Beskrivelse af Kong Christian 6. og Dronning Sophie Magdalenes Rejse til Norge 12. Maj–23. September,* facs. (Copenhagen: Poul Kristensens forlag, 1992). Reproduced with the permission of HM the Queen's Reference Library, Copenhagen.

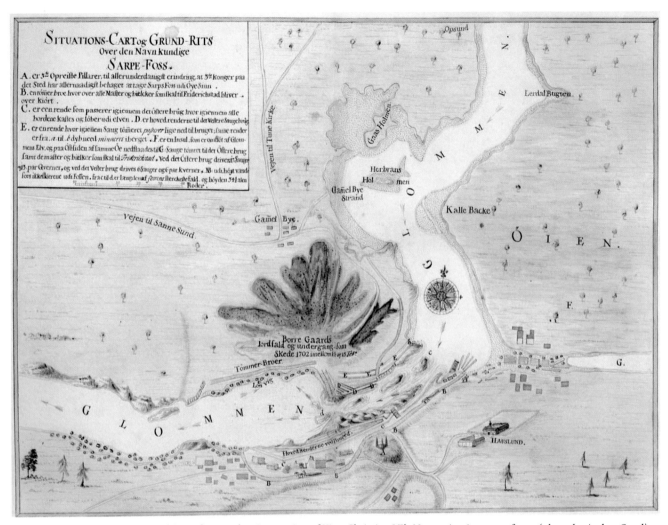

Figure 1.4 Map of the area around Sarpefossen, showing section of King Christian VI's Norwegian journey of 1733 (photo by Arthur Sand)
The letter A indicates the site from where three kings have viewed the waterfall. *Norske Reise Anno 1733. Beskrivelse af Kong Christian 6. og Dronning Sophie Magdalenes Rejse til Norge 12. Maj–23. September,* facs. (Copenhagen: Poul Kristensens forlag, 1992). Reproduced with the permission of HM the Queen's Reference Library, Copenhagen.

king, but the hearts of the Norwegians are even more so. After passing through the gates they could make progress to their quarters on a street flanked by spruce trees and armed citizens, all dressed in green. Cannons and arms were fired, and the citizens hailed the king and queen.

What is striking about the manuscript is the absence of any larger scale map outlining the Norwegian territory in full. The maps depict places where the king has been as well as principal towns, but there were also maps showing the roads in landscapes through which the king had travelled

Figure 1.5 Waarstigen, part of King Christian VIs Norwegian journey of 1733 (photo by Arthur Sand)

Norske Reise Anno 1733. Beskrivelse af Kong Christian 6. og Dronning Sophie Magdalenes Rejse til Norge 12. Maj–23. September, facs. (Copenhagen: Poul Kristensens forlag, 1992). Reproduced with the permission of HM the Queen's Reference Library, Copenhagen.

Figure 1.6 Manseidet, part of King Christian VI's Norwegian journey of 1733 (photo by Arthur Sand)

Norske Reise Anno 1733. Beskrivelse af Kong Christian 6. og Dronning Sophie Magdalenes Rejse til Norge 12. Maj–23. September, facs. (Copenhagen: Poul Kristensens forlag, 1992). Reproduced with the permission of HM the Queen's Reference Library, Copenhagen.

(Figure 1.4). No mountain is depicted without a road, and we see no road without the king and his company. In this sense nature is a theatre; the landscape is the backdrop, and the royal entourage the actors, a travelling company (Figure 1.5 and 1.6). The focus of the depiction is not the landscape itself but rather the route by which the king and his company move. On the opening page History inscribes the pyramid on the frontispiece of the manuscript, making a monument of the memory of the journey. This monument, the opening paragraph states, ought to be erected by the road at Dovrefjell, the most difficult passage on the tour. Memory as history could thus be inscribed in the landscape itself. Indirectly this is also what the large maps, and the maps in the album, do. They meticulously map out where the king has eaten and slept, as significant events surrounded by public ritual. The text on the maps marks the territory

where such ritual had taken place, annotating, thus, the presence of the king.

SCIENCE FOR THE KING

The commemorative manuscript must have been approved of by the king, as he soon gave the orders to have the manuscript printed. This was a long process, because of the lack of domestic artists and craftsmen. When in 1742 a Society for the Sciences was set up in Copenhagen, a society that was closely tied to and initiated by the king and his ministers, his first order was to have the manuscript printed. They were particularly looking for good etchings but also for new originals both to supplement and supplant the earlier drawings in the manuscript. Several draftsmen and etchers

Figure 1.7 Prospect de Güldelöw, Fridrichssteen, Storetaarn & Oberberg *(Mathias Blumenthal 1746, drawing on paper)*

Reproduced with permission of The Royal Danish Academy of Sciences and Letters.

worked towards publishing the tour in print, but in the end the plan was dropped. Some drawings and prints were made, but the whole manuscript was not printed until 1992. Since we have little evidence of the exact negotiations, the ambitions and intensions of the different parties are difficult to assess.[16] What is certain, however, is that none of those who participated in the process of making drawings and etchings took part in the journey. One of the most impressive etchings is an adaption of a drawing by the Italian professor in drawing at the Drawing and Painting Academy from 1740, Hieronimo Miani.[17] Comparing the drawing of the king's passage over the difficult mountain of Manseidet in the manuscript with the print that was made by Peter Cramer, the landscape has been turned into a mountain pass, Chinese style (Figure 1.8). The effects are dramatic, yet important symbolic features have disappeared, for example the spruce trees. Instead, Cramer's drawing equips the landscape with theatrical effects suitable to the theatricality of court life.[18]

Another solution of how to make drawings for commemorating the king's journey of 1733 was found by the painter Mathias Blumenthal, and in his case we can trace a shift in what counts as a good picture of nature. Probably already a resident of Fredrikshald in Southern Norway, he offered the Society of Sciences drawings of prospects, towns, churches and the like. In one painting we see the painter, Blumenthal, sitting in front of a framing device providing him with the perspective for depicting the city of Fredrikshald and the Fredriksten fortress. Paper, a measuring rod, a sheet of paper with a compass rose, and various other devices lie on the ground next to his feet. Three people eagerly follow his work while a fourth person is resting on the ground, observing the painter and his spectators. The middle ground of the painting shows stacks of wood and the harbour with different vessels. The background is densely filled with the city and the fortress of Fredrikshald. This rather unusual painterly solution is obviously based on his drawings made as an application to the Society of the Sciences, as we can see from drawing still in their archive (Figure 1.7).

What does the painter do in this painting? And what are the utensils doing here? Why does the artist focus on his own way of working? We do not know much about Blumenthal's exact motivations. He left few traces except for his paintings. One important aspiration, however, seems to be for the painter to emphasize his own presence in a more, conspicuous way than by simply depicting himself – as did so many painters in this period – as a viewer in his own work. His presence is the most important feature, forming the centre of the composition. In one of his drawings from roughly the same position, he adds in Latin 'as Nature shows'. Normally a prospect would help the viewer to know a place, the position of the fortress, the situation and quantity of the houses, the quality of the harbour, the most significant commercial activity. This painter insists that not only was he there, but he had the right tools for the job, and could work under the eyes of others, with witnesses.

Figure 1.8 Manseidet (etching by Peter Cramer, based on a drawing by Hieronimo Miani, n.d.)

The Italian drawing professor Miani made the original for this etching by Peter Cramer in a grand Chinese mountain style. 'Anno 1733 den 9 Augusti Reiste Kongelige Majestet Hendes Majestete Dronningen Samt Hendes Majestet Dronningens Fru Moder Over dette Field Mandseid/Miani Inventor P. Cramer Fecit'. Reproduced with the permission of The National Library of Norway, The Picture Collection.

What Blumenthal does is to direct our attention to the complex task of drawing up a landscape. As opposed to Miani and Cramer, he saw it as his task to produce drawings and paintings that were 'true to nature'. Such a description needed to be done in the right way, with the right tools. This was a different direction from what was considered good painterly style, as we can see from the contemporary text-book in drawing by Johan Preissler, where he states that the advantage in landscape painting is that it doesn't matter whether a tree is larger or smaller, a branch straight or curved; proportion is less important than in figure drawing.[19] As an application to the Scientific Society – not to the Drawing Academy – Blumenthals piece gives meaning. He chose differently from Miani and Cramer, and probably wisely so. His drawings included the king's loyal subjects, views of important fortifications and economic activities, all presented with scientific accuracy. In the minutes of the Society of Sciences it was explicitly stated in a letter to Blumenthal, that on the condition that the drawings were well done and that well-known men would attest to their accuracy, he would earn 16 Rdr for every drawing of Prospects, Places, Churches a.s.o in Norway.[20] The importance of witnessing in attesting to experiments has been documented as an integral part of the emergence and spread of the New Science from the seventeenth century.[21] In Blumenthal's painting the witnesses are included in the picture, attesting to the truthfulness to nature and as such securing the trustworthiness of a depiction that the receivers could not see for themselves.

Blumenthal was paid for delivering exact depictions, and his paintings differed significantly from other depictions we have seen from the king's tour in 1733. In these, nature was portrayed as a theatrical stage upon which the drama of the king's tour unfolded. In Blumenthal's drawings and paintings, on the other hand, nature delivers the blueprint, it is something fixed and given which is to be recorded faithfully. The difference is one of focus, even the figures in Blumenthal paintings are acting out a drama, their drama is about ascertaining the truth of the rendition of the landscape, the maps' dramas are about the king.

KNOWING AS NATURAL HISTORY

Coincidental with the work of the Scientific Society to publish the manuscript from Christian VI's Norwegian Tour, there emerged a new method for collecting information. Larger surveys of territory and populace had not been conducted before, but now this instrument of scientific investigation – and Royal government – was introduced through a list of 43 questions. They were issued by the

Gallerne En Farlig Vey under Filefield

Figure 1.9 Gallerne (etching based on a drawing by Mathias Blumenthal, photo by Arthur Sand)
Erich Pontoppidan, *Det første Forsøg paa Norges Naturlige Historie* (København, 1752/53). Private collection.

King's administration in Copenhagen and presented to all governors of counties and dioceses in Denmark, Island, the Faeroe Islands and Norway, in 1743. In particular it was the clergymen, the most widely dispersed and well educated group of state officials in the kingdom, who prepared the answers. But the questions were perhaps more important. They defined what was worth knowing about the King's territory. The size and shape of the territory, the animals and plants, the antiquities and archives, as well as the existence of gardens and the most common women's and men's names, were among the questions posed.[22] Observe, collect, describe, was the order issued from Copenhagen. The King demanded particular knowledge about his territory, not only about where he had been, but also about how his territory functioned as a unit. As in other European states, statistics were state secrets. The first publicly available information about the condition of the kingdom started to appear around the middle of the century. Perhaps one can see the attempt to present a printed version of the Norwegian progress as a first token of an opening up of the knowledge that was collected by the state. The output of this first survey from 1743, however, was meagre – that is, the answers flowed in to the central government, but the task of systematizing and applying this knowledge proved difficult.[23] This of course, is the great dilemma of collecting 'statistical' material, the amount of facts assembled makes it difficult to homogenize and utilize the results.

A more successful attempt at mapping the kingdom came from the industrious pietist ideologue, member of the Society of Sciences and priest to the King, Erich Pontoppidan, who arrived in Bergen as bishop in 1748. Well versed in the Bible and in civil and church history, he turned his eyes to nature. During his first three years as a bishop in Bergen he prepared a two-volume treatise, *Natural History of Norway*, which contained, according to its subtitle, cited here from the English edition which appeared in 1755: 'A particular and accurate Account of the Temperature of the Air, the different Soils, Waters, Vegetables, Metals, Minerals, Stones, Beasts, Birds, and Fishes; together with the Dispositions, Customs, and Manner of Living of the Inhabitants', and further, according to the subtitle, the book was 'Interspersed with Physiological Notes from eminent Writers, and Transactions of Academies'. This was an illustrated work of 800 pages about Norwegian nature.[24]

This is a work of physico-theology, a religious scientific textual practice and genre, but the book is also part of an attempt to make the knowledge of Norway travel to the king in Copenhagen. It is fundamentally descriptive, trying to encompass everything natural of importance. But the question is of course what it means to be fundamentally descriptive. The work depicted animals, birds and fishes, some cities and some scenes of agricultural work. It also presented some landscapes. The man who put his name to these was, again, Mathias Blumenthal, having moved to Bergen to work as a painter. But the landscapes he now draws have changed character. Here, he seems to use rather different pictorial conventions than in the drawings for the Scientific Society. The depiction of the mountain passage, Gallerne (Figure 1.9), has references in the texts where the difficult roads are treated quite extensively. The pictures are used to foreground particular aspects of the landscape, here the road, in others particular mountain formations.

Importantly this points to the power of conventions, to how the painter Blumenthal tries to convey the information about particular localities in the mode that suits a natural history best. Who actually asked him to deliver paintings in this style, is unknown. It has also been asserted that Blumenthal was unlikely to have visited the places he depicted and that he copied the drawings of others, which can of course account for the differences.

The 1743-survey and the natural history are important for visualizing the country, not because they contained many pictures, but because they took part in the emergent interest in mapping – and publicizing – resources. Norway was to be represented as more than rock and spruce. The human toil and industry brought to bear on the natural resources were starting to be regarded as the very basis for good government. To put it slightly differently, in line with English improvement theories and German cameralism, the government in Copenhagen got interested in the good governing of human and natural resources.[25] And Pontoppidan's natural history was a work directed at the government in Copenhagen.[26]

THE MOUNTAINS OF GOD

There is yet another feature of the *Natural History of Norway* that is of importance to the visual landscape depictions. It introduced an aesthetic vocabulary for talking about mountains and landscape. This was by no means exceptional, as Marjorie Hope Nicholson has shown in her book *Mountain Gloom and Mountain Glory* on this process in England. What she describes is the gradual transposition of the sublime from eternity to nature in late seventeenth and early eighteenth-century Britain and most notably in books that dealt with nature as the expression of God's wisdom, benevolence and might.[27] In a Norwegian context, Pontoppidan is the first to talk about mountains as a pleasurable sight instead of as a terrifying and difficult hindrance. When the author talks about the mountains in Norway, he talks about them as ornaments:

> Lastly, these natural fortifications seem also to be an ornament and decoration to the country; the diversified figures, and alternate eminences, and other varieties, according to the taste of most people, form a much more agreeable landscape than a flat and even country, which is almost every where the same. In this respect our country affords the most delightful contrasts in the diversity of its prospects. And these most magnificent structures of the great architect of nature, raise and animate the mind of man, by inspiring him with the most agreeable and the most sublime sentiments.[28]

Even if Norway can look terrible from the sea, people 'live very agreeably, amidst such delightful landscapes, that within a few miles, a painter might have choice of incomparable originals'.[29] Those who wanted more information about God, as the God of the mountains, were referred by Pontoppidan to the fourth chapter of William Derham's book *Physico Theology* from 1711.[30] For Pontoppidan the mountains were sublime and the landscapes delightful, because they were the works of God. Theologically and scientifically underpinned, Norwegian nature had become object for aesthetical consideration.

THE VIEW

After Pontoppidan's *Natural History*, many topographical treatises were written, dealing principally with smaller parts of the country, from parishes to counties. This became the most important literary genre of the Enlightenment in Norway. Written by clergymen and state officials, the treatesis described and collected natural resources of the local area. Many of the texts make for tedious reading, insisting that their project is merely to describe and collect what is there.

And what is there, is not views or landscapes. Instead, they describe the weather, the soil, the forests and fields, the principal economic activities and the manner of living of the inhabitants.

But there is at least one exception. The clergyman Jacob Nicolaj Wilse was one of the most important authors and editors in this genre, and his work about his own parish became a model for emulation. He was passionately interested in visual technologies, in mapping, in producing prints and landscapes.[31] He never failed to mention his *camera obscura*, or his methods of mapping, and in his texts, he presents his reader with numerous views. These are moments when landscapes suddenly appear as worthy of being painted – rare moments in the topographical literature. Where other authors describe the distance between one river and the next, one mountain and the other, Wilse leads the reader on walks to hilltops to get overviews, to experience 'the situation' through his text. The reader has to follow him, maybe to experience and get a feeling for the territory, maybe to better know the territory. He also includes plates with views in his topographies. In his *Physical, economic and statistic description of Spydeberg* from 1779, Wilse has included one plate with seven windows, as he calls them.[32] He made them himself with his *camera obscura* to make the first impressions from nature. Afterwards, a draftsman would treat them artistically, Wilse informs the observer in the accompanying text. But what did the view do in this kind of collection, whose aim was to describe local things, not to consider place as a visual experience?

ON THE ROAD AGAIN

In 1788 a new regent toured Norway. Crown Prince Frederik was the *de facto* ruler of the state, and he organized an extensive tour in the footsteps of his ancestors. He also returned briefly the same year to fight a war against Sweden. The artist Erik Pauelsen followed in the wake of the king, making pictures and views. Pauelsen was not interested in depicting the king, nor the road. His motifs were the views from the road. The pictures were intended to form a first *Voyage Pittoresque* of Norway, in the Swiss style, with Charles Melchior Descourtis as a model.[33] Unfortunately Pauelsen died and the project was abandoned, but an accompanying text was printed in 1790 under the title 'About Norway'. Here it is stated that the topographic descriptions are so dry, that they make it impossible to acquire a true conception of the picturesque appearance of the country. This is all the more a shame, according to the author Christen Pram, since what the appearance of the country can convey to us is the quality of government and the nature of the political system. 'Apart from the political beauties which are the foundations for patriotism, and which delight every philantrope, there also exist more or less physical painterly beauties, [...] which leads to patriotism and delight any onlooker'.[34] In the same vein the introduction to the *Topographical Journal* [*Topographisk Journal*] in 1792, expressed how topography brought pictures of the political and economical beauty of a country, a beauty dependent on the political freedom that reigned in the country.[35] Switzerland was the country to identify with, politically as well as aesthetically, because 'inhabitants of mountainous countries are always the most patriotic'.[36]

The feeling for nature was a political and metaphysical/religious feeling. And here it is time to go back to where I started: *The artistic discovery of Norway*. The author of the book, Carl Schnitler, described how art was becoming a means of expressing the needs of the soul. It was becoming a conscious life-power and an important ally for the basic values in the coming century: the feeling for nature and the idea of the nation. This was the artistic discovery of Norway in 1789, according to Schnitler. Thus the king (or more precisely prince) on the road was followed by an artist who had acquired a feeling for nature. But this feeling for nature was by no means apolitical, it was harnessed both by the king and those who worked for promoting a more republican Norwegian patriotism. Theology, aesthetics, politics, and

resource management had made imprints on Norwegian nature, and the effect was more than one feeling for nature.

ENDNOTES

1   Carl W. Schnitler, *Norges kunstneriske opdagelse. Maleren Erik Pauelsens norske landskaper 1788* (Kristiania: Steenske forl., 1920).

2   The 'supplikk' was printed in Christiania and contained a verse in honour of the King in Latin and Danish. At the very bottom a few lines describe the occasion for the request, the lack of funds for the parish church in Sel. Archives of *De Sandvigske Samlinger*, Maihaugen, SS-DS-1995-0022.

3   Jacques Revel, 'Knowledge of the Territory', *Science in Context* 4/1 (1991), pp. 133–161.

4   Ovid, *Metamorphoses*, trans. Georg Sandys, Oxford, 1632, facs. (Lincoln: University of Nebraska Press, 1970), p. 458.

5   The best overview is to be found in Anne-Mette Nielsen, *Kongeferder i Norge gjennom 300 år* (Lillehammer: Norsk Vegmuseum, 1999).

6   The 1749 tour was a visitation journey, but the king was unable to move much beyond Christiania, and the tour is not discussed in Nielsen, ibid.

7   The best introduction to early modern landscape and folk-life in paintings is Leif Østby, *Med kunstnarauge. Norsk natur og folkeliv i biletkunsten* (Oslo: Det norske Samlaget, 1969).

8   Ibid.

9   H. Ehrencron-Müller, 'Et dansk pragtværks historie', *Nordisk tidskrift för bok- och biblioteksväsen* (1917), pp. 313–327.

10 See *Norske Reise Anno 1733. Beskrivelse af Kong Christian 6. og Dronning Sophie Magdalenes Rejse til Norge 12. Maj–23. September*, facs. (Copenhagen: Poul Kristensens forlag, 1992); Ehrencron-Müller, 'Et dansk pragtværks historie'; Asger Lomholt, 'Christian VI's Rejse i Norge 1733'

in *Det Kongelige Danske Videnskabernes Selskab 1742–1942. Samlinger til Selskabets Historie* (København: Ejnar Munksgaard, 1960).

11 Ehrencron-Müller, 'Et dansk pragtværks'; Lomholt, 'Christian VI's Rejse'.

12 'Reise Carte over Deris Kong. Mayts og Hendis Mayt Dronningens Tour i Norge til Lands hvor paa Weyernes Længde, Elver og Broer efter Mastaben udi Grund-Rits samt de dificileste Passager over Klipperne udi Perspective Ritser forestilles 1733'.

13 Francesca Fiorani, *The Marvel of Maps. Art, Cartography and Politics in Renaissance Italy* (New Haven: Yale University Press 2005), p. 11.

14 In 'Christian VI's Rejse i Norge 1733' it is assumed that he was a copist in the Danish Cancelli who died in 1760 as a Magisterpresident in Bergen. One does not know who made the original drawings.

15 See Edward Muir, *Ritual in Early Modern Europe* (Cambridge: Cambridge University Press, 1997).

16 See Lomholt, 'Christian VI's Rejse'.

17 Lomholt, 'Christian VI's Rejse', pp. 57–8.

18 Cf. Kenneth Olwig, *Landscape, Nature and the Body Politic* (Madison: University of Wisconsin Press, 2002).

19 Johann Daniel Preissler, *Gründliche Anleitung welcher man sich im Nachzeichnen schöner Landscafften oder Prospecten bedienen kan, den Liebhabern der Zeichen=Künst mitgetheilet und eigenhändig in Kupffer gebracht* (3rd edition, Nürnberg, 1746).

20 Ehrencron-Müller, 'Et dansk pragtværks', p. 323.

21 Authoritative in this huge literature is Steven Shapin, *A Social History of Truth: Civility and Science in Seventeenth-Century England* (Chicago: University of Chicago Press 1994).

22 Kristin Røgeberg, *Norge i 1743*, vols 1–5 (Oslo: Solum forlag, 2003–8).

23 Erich Johan Jessen-Schardebøll was responsible for collecting the information and nine volumes was planned, but eventually only one book appeared 20 years later: Jessen-Schardebøll. *Det Kongerige Norge fremstillet efter dets naturlige og borgerlige tilstand 1* (København 1763). Cf. ibid, p. 41.

24 Erik Pontoppidan, *Det første Forsøg på Norges Naturlige Historie*, vols 1–2 (Kiøbenhavn 1752–3), translated to English as *The Natural History of Norway* (1755).

25 See Lisbet Koerner, *Linnaeus: Nature and Nation* (Cambridge, Massachusetts: Harvard University Press, 1999) for a discussion of Scandinavian cameralism, a German/Scandinavian cousin of mercantilism. On improvement, see also Finola O'Kane in this volume.

26 The volumes were prefaced by dedications to two of the five ministers of the king. See Pontoppidan, *Natural History*.

27 Marjorie Hope Nicolson, *Mountain Gloom and Mountain Glory* (Seattle: University of Washington Press, 1997 [1959]).

28 Pontoppidan, *Natural History*, p. 64.

29 Ibid.

30 Ibid., p. 65.

31 See especially *Physisk, oeconomisk og statistisk Beskrivelse over Spydeberg* (Christiania 1779); *Reise-Iagttagelser i nogle af de nordiske Lande*, vol. 1 (Kiøbenhavn 1790); *Topographisk Journal* (1792).

32 Wilse, *Physisk, oeconomisk*, p. 22.

33 Schnitler, *The Artistic Discovery*, p. 24.

34 Christen Pram, 'Om Norge' i *Minerva* 1790, reprinted in Schnitler, *The Artistic Discovery*, p. 60.

35 'Om Topographisk Journal' in *Topographish Journal*, 1792, no. 1, vol. 1.

36 Pram, 'Om Norge', p. 63.

2 Flow: Rivers, Roads, Routes and Cartographies of Leisure

Vittoria Di Palma

In so variegated a country, as England, there are few parts, which do not afford many pleasing, and picturesque scenes. The most probable way of finding them, as I observed a little above, is to follow the course of the rivers. About their banks we shall generally find the richest scenery, which the country can produce.[1]

One morning in the spring of 1764 William Gilpin and his younger brother Sawrey set out on a trip down the River Thames. They hired a boat at Windsor and spent the day sailing downstream to London, Gilpin recording his observations in a small notebook, Sawrey making sketches of the scenes they encountered along the way. Their aim, as William Gilpin was to recount many years later, was to survey the river along its length in order to create a volume of river scenery. His plan was 'to navigate the river, and fix the principal points on it's [*sic.*] banks. These would remain as a kind of landmarks, from which the appearance of the river from the land might afterwards be examined, and anything remarkable in it's [*sic.*] neighbourhood might be brought within a picturesque survey'.[2] This survey was to begin at Oxford, the place at which the Thames became a 'river of consequence', and continue downstream until it reached the sea.[3] In order to make the project more feasible, the Thames was to be surveyed in parts, and to that end Gilpin divided it into three sections, each one marked by a distinctive character. In the first section, from Oxford to Windsor, the Thames mostly passes 'through meadow-lands, and scenes of cultivation', and its character is primarily rural.[4] In the second, from Windsor to London, the villas adorning its banks, and the pleasure boats plying its reaches give the river 'an air of high improvement, expence, and splendor'.[5] Finally, in the third section, from London to the sea, the Thames's

broad, bleak shores [...] are forsaken by the gay villas, which stood thick upon its higher stream; and its vast oozy waters, having now lost all power of reflecting objects, and indeed having nothing on their banks worth reflecting, glide only as the tide ebbs and flows, past naked villages, and smoaky fish towns. Its light skiffs are changed into large ships, both of force, and burden; and it becomes intirely a naval, and commercial river.[6]

Gilpin's ambitious project, however, was never completed. Although it was England's principal river, of paramount importance both historically and geographically, the Thames was disappointing from a picturesque point of view: its banks too low, its wood too thinly scattered, and its villas too ornamented for Gilpin's taste. In the end, only the section from Windsor to London was navigated and surveyed, and Gilpin's notes and Sawrey's sketches were not tidied up and made into a clean manuscript copy until many years later.

It is probable that the Thames's aesthetic inadequacies were largely responsible for the abandonment of this particular project, for Gilpin continued to be captivated by rivers and the aesthetic experience they could provide for the rest of his life. In fact, Gilpin found river scenery so varied and aesthetically complex that he believed a survey of England's rivers could generate a complete compendium of its landscape. 'I have often thought', Gilpin wrote in *Observations on Several Parts of England, particularly the Mountains and Lakes of Cumberland and Westmoreland, Relative Chiefly to Picturesque Beauty* of 1786,

> *that if a person wished particularly to amuse himself with picturesque scenes, the best method he could take, would be to place before him a good map of England; and to settle in his head the course of all the chief rivers of the country. These rivers should be the great directing lines of his excursions. On their banks he would be sure, not only to find the most beautiful views of the country; but would also obtain a compleat system of every kind of landscape.[7]*

This grand aim of navigating and charting all of England's rivers was only ever realized piecemeal, and although fragments of river surveys occur in a number of Gilpin's works, and numerous drawings of rivers can be found in both his manuscripts and printed works, it is his *Observations on the River Wye, and Several parts of South Wales, &c. Relative Chiefly to Picturesque Beauty: Made in the Summer of the Year 1770*, but not published until 1782, which contains his most sustained consideration of rivers and their settings.[8] Gilpin's version of the picturesque is often caricatured as a practice that stilled the motion and life of landscape into a series of static, framed views. However, by examining his analysis of fluvial scenery it becomes apparent not only that Gilpin was extremely attentive to the experience and effects of motion, but also, and more importantly, that his analysis of rivers was key to his formulation of an aesthetics of a mobile eye on a winding route.

RIVER ROUTES

Although Gilpin's picturesque goal of obtaining 'a compleat system of every kind of landscape', was novel, his method of taking rivers as a pattern for his excursions had a venerable history. Gilpin himself cites the precedent of Polidore Vergil's *Anglica Historia* of the early sixteenth century, which includes a description of the Thames from source to mouth, but this was just one of a number of chorographical accounts that used rivers as a basis for organizing information. Two other important precedents were William Harrison's 'Description of Britain', published as part of Raphael Holinshed's *Chronicles* of 1577, which included a long section on the nation's principal waterways and, even more importantly, William Camden's magisterial *Britannia*, first published in Latin in 1586, which used the course of local rivers to order its county descriptions'.[9] More than merely a utilitarian system of inland navigation, rivers were a symbol of the country's very health and vitality. As Sir Edward Turner, Speaker of the House of Commons, said in his speech to a joint session of the Houses of Parliament in 1665, 'Cosmographers do agree that this Island is incomparably furnished with pleasant Rivers, like Veins in the Natural Body, which conveys the Blood into all the Parts, whereby the whole is nourished, and made useful'.[10]

Rivers may have been powerful symbols of circulation and connection, and river routes may have provided a compelling pattern for the organization of literary works, but in practice their navigation was notoriously cumbersome.[11] In England rivers were governed by laws that dated back to the Magna Carta, which differentiated between tidal and non-tidal waterways. Tidal rivers belonged to the Crown, and were, like public highways, free for common use. Non-tidal rivers, on the other hand, were understood to be private property, belonging to the landowners who could lay claim to their banks.[12] As water depth in many rivers tended to be notoriously inconsistent – affected by the season, by tidal ebbs and flows, and by the presence of discarded ballast or garbage that could cause boats to run aground – landowners (millers in particular) made a practice of constructing elaborate systems of dams, weirs, and locks in the portions of the rivers they controlled to regulate its flow. These, along with other impediments like fish garths, not only obstructed the river, but often made an already shallow stretch of water even shallower. In these shallow sections, barges were dependent on being granted what was called a 'flash' or 'flush' of water to float them over, putting boat and barge drivers at the mercy of millers who could charge extravagant fees or refuse the privilege altogether, leading to delays that could last hours or even days. Finally, vessels for hire often charged high (some said extortionate) fees, and watermen were notorious for their incompetent boatmanship, as well as for their rough manners and colourful language that could offend those passengers of a daintier sensibility.[13] Notwithstanding the difficulties, dangers, and expense associated with travel by land, in England travel by river was the exception rather than the rule.

In fact, the history of river navigation in England is the story of a perpetual struggle between competing interests. Cities whose economies depended on rivers for the transportation of their goods and provisions tended to be in favour of improvement schemes calling for the removal of dams and weirs and the deepening of channels. Millers, on the other hand, needed dams for the operation of their mills, and landowners whose holdings abutted a river argued that removing dams would lead to flooding that would ruin crops and decrease the value of their estates. Rivers may have been the veins of the nation, carrying its lifeblood into all parts, but as schemes for their improvement foundered on the shoals of entrenched interests, it became increasingly clear that alternatives for mercantile transportation needed to be found. Although the first English canal is usually identified as that built by the municipal authorities of Exeter in 1564, the era of canal building did not really begin in earnest until 1760, marked by the construction of James Brindley's innovative Bridgewater Canal for the 3rd Duke of Bridgewater's coal mines at Worsley. This was followed by the Grand Trunk

Canal in 1766, the Staffordshire and Worcestershire canal in 1772, the Stroudwater Canal in 1777, and, between 1795 and 1805, the engineering triumph of Thomas Telford's and William Jessop's Pontcysyllte Aqueduct, which carried the Llangollen Canal high above the valley of the River Dee in Wales. By 1820, approximately 3,000 miles of canal had been constructed in England and Wales, utterly transforming the country's inland navigation and irrevocably altering its countryside.[14]

For much of its history, the River Wye, on the border between England and Wales, was caught in a struggle that conformed to a common pattern. In the time of Edward I (1239–1307) the Wye had been categorized as a public navigable river, but in the succeeding centuries its stream had gradually become encumbered by the erection of numerous dams and weirs. Over the course of the sixteenth and seventeenth centuries a series of Acts of Parliament were passed calling for their removal, but it was not until 1662 that Sir William Sandys was commissioned to improve the Wye's navigation by dismantling the existing weirs and installing a series of locks. Unfortunately, his methods, which had worked so well on the sluggish River Avon, were not suited to the more sprightly Wye, and it was not until 1695, when a new bill was passed that authorized the demolition of all dams and weirs (with the exception of the New Weir near Goodrich), that the river's navigation was markedly improved.[15] However, according to the Reverend Thomas Dudley Fosbrooke, author of *British Monachism, or, Manners and Customs of the Monks and Nuns of England* as well as a popular guidebook of the Wye and its surrounding region, even as late as 1818 'the term *navigable*, applied to the Wye, merely denotes a depth of water, sufficient at certain seasons, for the passage of barges. The river is not a majestic flowing stream, roiling its waves along in grand unceasing procession; but in winter, a bustling hurrying torrent; in summer, a broad rivulet lounging carelessly through a rocky ravine'.[16] Thus, the 'broad rivulet' that William Gilpin set out to chronicle in the summer of 1770 and immortalized in print with the publication of his *Observations on the River Wye* 13 years later was not a waterway with a long and established history of fluent navigation. Although Gilpin was certainly not the first person to sail down the Wye for recreational rather than mercantile purposes, it was unquestionably his book that established the Wye and its surrounding landscape as a destination, setting the pattern for others to follow, and creating, in this way, the touristic experience that was subsequently to be known as 'The Wye Tour'.[17] Gilpin's decision that *Observations on the River Wye* would be the first of his tours to be published also forged an enduring link between rivers and tourism in general. For as canals began to play the leading part in the nation's commercial transportation network, rivers were free to take on a new role as generators of an infrastructure of leisure.

Figure 2.1 William Gilpin, Observations on the River Wye, and Several Parts of South Wales, &c. Relative Chiefly to Picturesque Beauty; Made In the Summer of the Year 1770 *(London, 1782) (photo by Daniel Talesnik and Nicolas Stutzin)*

Reproduced with permission of Collection Robin Middleton.

FLUVIAL AESTHETICS

In June of 1770, Gilpin set out from Cheam, Surrey, and headed west. When he reached Ross, on the banks of the Wye, he hired a covered boat and three men to row it; his plan was to travel downstream from Ross to Monmouth, and then to continue his journey by land after reaching the point where the river became too shallow and rapid for the barge to proceed.[18] Although Gilpin had undertaken ambitious journeys before – most notably around Kent in 1768, and to Essex, Suffolk, and Norfolk to see Lord Orford's collection of pictures at Houghton in 1769 – his tour of the Wye was conceived differently: it was the first to have the aim 'of not barely examining the face of a country; but of examining it by the rules of picturesque beauty; that of not merely describing; but of adapting the descriptions of natural scenery to the principles of artificial landscape; and of opening the sources of those pleasures, which are derived from the comparison'.[19] For this purpose, the River Wye was eminently adapted. Its lofty banks and mazy course resulted in landscapes that were 'the glory of river scenery', and its views were favoured with

'the most elegant kind of perspective; free from the formality of lines' (Figure 2.1).[20]

These views contrasted sharply with those offered by canals. Gilpin began his tours just as Britain's great (though short-lived) era of canal-building was getting underway, and he was appalled by the way in which these 'cuts', as he called them, deformed the landscapes they passed through. Canals were artificial rivers, condemned by Gilpin for their dissimulation: 'an object, disgusting in itself, is still more so, when it reminds you, by some distant resemblance, of something beautiful'.[21] Whereas a noble river 'winding through a country; and discovering its mazy course, sometime half-concealed by it's woody banks and sometimes displaying it's ample folds through the open vale' was 'one of the most beautiful objects in nature', a canal was its opposite in every respect: 'it's lineal, and angular course – it's relinquishing the declivities of the country; and passing over hill, and dale; sometimes banked up on one side, and sometimes on both – it's sharp, parallel edges, naked, and unadorned – all contribute to place it in the strongest contrast with the river'.[22] River views were made up of four principal parts – 'the *area*, which is the river itself; the *two side-screens*, which are the opposite banks, and mark the perspective; and the *front-screen*, which points out the winding of the river', but the linearity of canals meant that the views they offered had 'no front-screen: the two side-screens [...] lengthen to a point'.[23] In other words, while canal views were governed by the rules of one-point perspective, the views offered by a sinuous or serpentine river were free of those strictures, embodying instead the aesthetics of the picturesque.

The serpentine line – William Hogarth's 'line of beauty' – was emblematic of nature, and nature's liberty.[24] According to Thomas Whately, author of the enormously influential *Observations on Modern Gardening* and Gilpin's friend and neighbour, the serpentine line refused to be subjugated, keeping itself 'at a distance from every figure, which a rule can determine, or compass describe'.[25] The serpentine line was natural, free, and therefore beautiful: Hogarth found it to be the most ornamental of all lines because it was the most varied: 'by its waving and winding at the same time different ways, [it] leads the eye in a pleasing manner along the continuity of its variety [...]'.[26] But above all, the serpentine line was inextricably bound to the concept of motion. Hogarth remarked on the way that 'the hand makes a lively movement in making it with pen or pencil', and Henry Home, Lord Kames, articulated a common association between the serpentine line, freedom, and motion when he wrote in his *Elements of Criticism* that rather than liking motion in a straight line 'we prefer undulating motion, as of waves, of a flame, of a ship under sail. Such motion is more free, and also more natural. Hence the beauty of a serpentine river'.[27] Motion was also a quality inherent to water itself, no matter where it was located, or what form it took. As Gilpin observed, 'on the *plain* it rolls majestically along, in the form of a deep-winding river. In a *mountainous country* it becomes sometimes a lake, sometimes a furious torrent broken among shelves and rocks; or it precipitates itself in some headlong cascade. Again, when it goes to sea, it sometimes covers half a hemisphere with molten glass; or it rolls about in awful swells; and when

it approaches the shore it breaks gently into curling wave, or dashes itself into foam against opposing promontories'.[28] For Gilpin and his contemporaries, therefore, the River Wye would have been a natural entity that embodied the idea of motion both in its serpentine form and in the constant flow of its watery substance.

But the Wye did not just incarnate the idea of motion, for it was not simply an object to be looked at. Even more than an object, the Wye was an experience, something that could only fully be apprehended through one's movement along it. And if the Wye's 'mazy course' provided one facet of its aesthetic appeal, its 'lofty banks' added another. The combination of a serpentine course and high wooded banks produced views that were discrete and bounded, visible one at a time, one after another. Gilpin recounts that as his party first set out from Ross:

we met with nothing, for some time, during our voyage, but these grand woody banks, one rising behind another; appearing, and vanishing, by turns, as we doubled the several capes. But though no particular objects marked and characterized these different scenes; yet they afforded great variety of beautiful perspective views, as we wound round them; or stretched through the reaches, which they marked out along the river.[29]

Gilpin's description gives eloquent voice to the effect of motion on visual experience. Following the course of the river gives rise to a sequence of encounters with scenes that appear, advance, and vanish by turns. But Gilpin is not describing a series of absolutely distinct and isolated scenes, or a sequence of static, framed views. Instead, his text suggests a continuity between one scene and the next – a continuity that is produced by Gilpin's motion along the river. Furthermore, although Gilpin is of course aware that he and his boat are in motion, winding along the river's capes and gliding along its stretches, instead he experiences the landscape as though *it* is moving towards *him*.

Figure 2.2 William Gilpin, Observations on the River Wye, and Several Parts of South Wales, &c. Relative Chiefly to Picturesque Beauty; Made In the Summer of the Year 1770 *(London, 1782) (photo by Daniel Talesnik and Nicolas Stutzin)*

Reproduced with permission of Collection Robin Middleton.

Gilpin's *Observations on the River Wye* is an essay on the aesthetics of landscape and motion. And even though a river may have provided a particularly apt case study for the analysis of how landscape is apprehended by a mobile eye on a winding route, a road was not a bad substitute (Figure 2.2). In fact, Gilpin's most sustained evocation of landscape experienced in motion occurs in his description of a stretch of road near Dryslwyn Castle, in Carmarthenshire, well worth quoting in full:

> *Having thus passed the mount Cenis[30] of this country, we fell into the same kind of beautiful scenery on this side of it, which we had left on the other: only here the scene was continually shifting, as if by magical interposition.*
>
> *We were first presented with a view of a deep, woody glen, lying below us; which the eye could not penetrate, resting only on the tops, and tufting of the trees.*
>
> *This suddenly vanished, and a grand, rocky bank arose in front; richly adorned with wood.*

> *It was instantly gone; and we were shut up in a close, woody lane.*
>
> *In a moment, the lane opened on the right, and we had a view of an inchanting vale.*
>
> *We caught its beauties as a vision only. In an instant, they fled; and in their room arose two bold woody promontories. We could just discover between them, as they floated past, a creek, or the mouth of a river, or a channel of the sea; we knew not what it was: but it seemed divided by a stretch of land of dingy hue; which appeared like a sand-bank.*
>
> *This scene shifting, immediately arose, on our left, a vast hill, covered with wood; through which, here and there, projected huge masses of rock.*
>
> *In a few moments it vanished, and a grove of trees suddenly shot up in its room.*
>
> *But before we could even discover of what species they were, the rocky hill, which had just appeared on the left, winding rapidly round, presented itself full in front. It had now acquired a more tremendous form. The wood, which had before hid its terrors, was now gone; and the rocks were all left, in their native wildness, everywhere bursting from the soil.*
>
> *Many of the objects, which had floated so rapidly past us, if we had had time to examine them, would have given us sublime, and beautiful hints in landscape: some of them seemed even well combined, and ready prepared for the pencil: but, in so quick a succession, one blotted out another. – The country at length giving way on both sides, a view opened, which suffered the eye to rest upon it.[31]*

Gilpin's account of his headlong descent is a verbal exercise in the picturesque mode, a narration of his encounter with a variety of scenes, each one individual and distinct – here conveyed by the way each paragraph is framed and set off by the white space of the page – but joined one to another in sequence. Motion is the connecting force, evoked by phrases such as 'this suddenly vanished', 'it was instantly gone', or 'this scene shifting', which give a sense of objects appearing, looming, and passing by, changing position, shape, and appearance as they do so.

Gilpin's text does not present an intellectual conception of motion, but rather its effect: he describes his descent as though he is the passive spectator of a landscape that is hurtling by – a characterization that vividly evokes Kames's formulation of the cognitive process. For Kames, motion is what makes thought possible. Thought is a series of perceptions and ideas, a continuous train of objects over which we have no control: 'It requires no activity on [an individual's] part to carry on the train nor has he power to vary it by calling up an object at will', he writes in a chapter entitled 'Perceptions and Ideas in a Train'.[32] For Kames, ideas are presented to the mind and connected one with another precisely as objects are externally, making external relations the pattern for internal ones. In fact, his construct of the thinking subject is uncannily similar to the figure of a traveller, seated, perhaps in a boat or carriage, and mesmerized by the landscape fleeting by.

Gilpin's attempt to create an aesthetics of landscape that centres around the mobile subject applies Kames's theories to the visual realm. For Gilpin, landscape was not merely a collection of objects to be stilled and gathered into a frame, but rather a series of views bound one to the next by a viewer's motion. Gilpin's tours, both in manuscript and when they were eventually printed, are illustrated by sequences of views – each one distinct, but properly understood only as part of a series. Each view is taken from the point of view of a traveller in a landscape, and many scenes even include one or two small wayfarers to provide a point of identification for the viewer as well as a sense of scale. The views often make use of the compositional device of a winding river or road stretching into the background to generate a sense of depth and encourage

the viewer's virtual motion into the scene, and at times, when viewed in sequence, it is clear that particular sets of views depict a single place from different vantage points, evoking the changing aspect of an object or scene as one approaches, arrives, and passes by. The views are usually oval in shape, and tinted in such a way to recall an image seen in a Claude Glass – a device that framed but most certainly did not still the landscapes it reflected. Gilpin's description of the experience of using one to look at the landscape outside the window of his chaise gives voice to an aesthetic capable of comprehending both an individual scene, and a visual sequence:

[...] a succession of high-coloured pictures is continually gliding before the eye. They are like the visions of the imagination; or the brilliant landscapes of a dream. Forms, and colours, in brightest array, fleet before us; and if the transient glance of a good composition happen to unite with them, we should give any price to fix, and appropriate the scene.[33]

Gilpin's books on the picturesque beauties of England, Scotland, and Wales created a vogue for domestic tourism that privileged certain kinds of landscapes and established the pattern for viewing them. They encouraged the appreciation of scenery from a mobile point of view, schooling aesthetically-minded travellers in the principles of picturesque beauty and the aesthetics of landscape in motion. But although Gilpin's concept of picturesque tourism may have been new, his method of generating an understanding of the national landscape in terms of travel through it had a longer history, one that was closely bound up with cartographic innovations of the previous century.

ROADS AND ROUTE MAPS

A river presents the traveller with a particular kind of course, one that combines characteristics of both road and route.

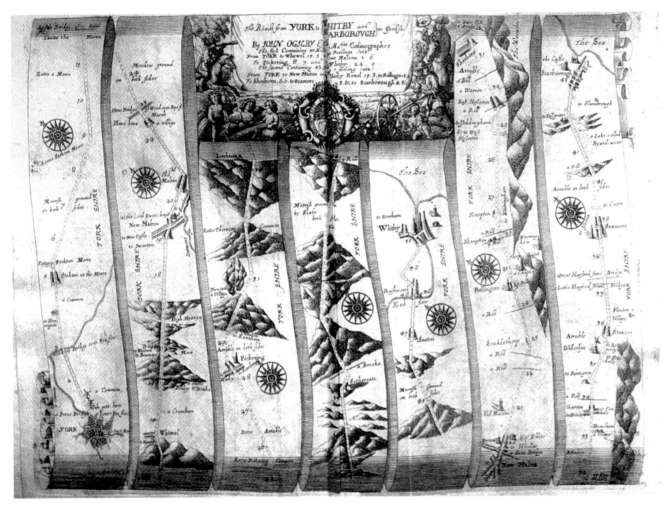

Figure 2.3 John Ogilby, Britannia, Volume the First: Or, an Illustration of the Kingdom of England and Dominion of Wales: By a Geographical and Historical Description of the Principal Roads thereof *(London, 1675)*

Reproduced with permission of The British Library Board, Maps C.7.d.8.

A road is a physical entity, what Gilpin would have called 'an object' in a landscape. A route, on the other hand, is an abstraction: destination-driven, it projects or records the way from A to B.[34] A road can be travelled along from beginning to end, or left at various points. Whether one follows it or not, a road continues to exist as an entity. Together with other roads, it forms a network. A route, on the other hand, as a line that connects two places, is more idea than object. It has a beginning and an end, it presents a series of directions arranged in sequence, and whether it is formulated on paper, in an oral utterance, or in a traveller's mind, it exists only insofar that it is followed. Although a river, like a road, is an object in the landscape, it is not a constructed entity, but a natural one. Like a route, a river has a beginning (its source) and an end its (mouth), and to travel along it implies sticking to its course, but its sequence is a given, established by nature rather than human needs. Nevertheless, floating downstream (preferable to rowing upstream, but it's the same no matter the direction) entails coming upon places, objects, scenes, and landscapes in a particular order. To travel the course of a river is in many respects like following an itinerary, involving a sequence of predetermined encounters with items on a list.

As a form of wayfinding guide, the itinerary had a long and venerable history in England, stretching all the way back to the *Iter Britannicum* or the British portion of the fourth-century Antonine Itinerary, the register of the routes connecting the major settlements, camps, and forts of the Roman Empire.[35] Around the beginning of the sixteenth century, types of wayfaring aids that undoubtedly had previously existed in manuscript, such as itineraries, lists of roads, and distance tables, began to be published, and the shift from manuscript to print led not only to greater availability but also to increased standardization. England was crisscrossed by a network of thoroughfares of differing size and importance, from major highways to the smaller and more local ways, paths, and tracks whose various combinations could offer a multitude of ways to get from one place to another. But once itineraries began to be printed, and as the country's postal system developed, particular sequences of roads, punctuated by designated inns and changing places, began to congeal into routes. The term 'way' became specific and directional, indicating a course to be followed ('the way to x'), rather than simply a path.

Up until the middle of the seventeenth century, printed itineraries tended to be small, cheap, and portable. Designed to be carried in a pocket or bag to help with wayfinding *en route*, for the most part they were lists or tables of locations and distances, sometimes including information about markets and fair days. But in 1675 a publication appeared that profoundly affected the subsequent form and history of the genre: John Ogilby's lavish folio, *Britannia.* Dedicated to Charles II, *Britannia* was a pioneering work that aimed

to survey and map over 40,000 miles of the kingdom's post roads.[36] Its 100 maps helped to standardize the mile at 1,760 yards,[37] established the scale of one inch to the mile, and introduced a striking and innovative graphic convention for the representation of itineraries, one with important consequences for the way travel was to be envisioned in England for the next century and beyond: each route is represented as though drawn on a continuous strip or scroll of paper that loops and doubles back upon itself as it unfurls from left to right across the space of the page (Figure 2.3).[38] At the beginning of each scroll, located at the lower left of each plate, is the city of origin – usually London – which is represented ichnographically. As the route continues through the countryside, it fords rivers, traverses heaths, passes through hamlets, crosses bridges, and ascends and descends mountains until it reaches its city of destination – usually also represented in plan – located at the top right of the same or a successive plate.

Ogilby's strip maps – descendents of the Roman Peutinger Table and precursors of contemporary navigational aids like Google Maps or the American Automobile Association's TripTiks – are pictorial itineraries, charting routes rather than roads.[39] Not only do they give visual form to the abstraction of a route, but also, and more importantly, they generate an image of the landscapes passed through. They use words and icons to denote features and objects on or to either side of the road including cities, villages, and hamlets; churches, windmills, houses of the gentry and bridges of wood and stone; mountains, rivers, and seas. They indicate whether the roads are open or enclosed by hedges or stone walls; describe the types of terrain traversed, whether arable, pasture, or forest, heathy, boggy, or moorish; and make note of the presence of natural resources such as iron or coal deposits. To do this the maps make use of the cartographic convention of combining ichnographic and perspectival views – larger towns, rivers, lakes, and the road itself are shown in plan or as though seen from above, while objects like trees, villages, churches, mills, houses of the gentry, and the occasional castle are depicted in elevation. This convention is particularly striking when applied to the representation of hills and mountains which Ogilby depicted, possibly on the advice of Robert Hooke, in elevation and right side up, to indicate an ascent, and in elevation but upside down, to denote a descent.

Ogilby's strip maps established a particular way of visualizing travel, one that united the list-like quality of the itinerary with a pictorial interest in representing local landscapes. A strip map, by definition, is concerned almost exclusively with the terrain to either side of its chosen route, sacrificing consistent orientation and a sense of a greater topographical context for more proximate information. In Ogilby's maps, the objects depicted to either side of the route, whether towns or individual buildings, mountains, woods, or fields of grain, are arranged along a line like beads on a string. To read the map involves moving one's eye from bottom to top, left to right, and noting each of these objects in sequence. To travel while using a map configured in this way would mean moving through the landscape attuned to a particular succession of objects, noting each one as it was encountered, anticipating the next one lying just ahead.

Although *Britannia*'s size and heft may have made it unlikely that the cumbersome folio was ever used for wayfinding *en route*, it unquestionably set the pattern for British route maps for the next century and beyond, spawning an unruly brood of reprints, epitomes, knock-offs, and imitations.[40] That *Britannia* established what Ogilby called the 'Itinerary Way' and its strip-map incarnation as the template for British travel becomes evident by taking an overview of the road books published in England during the following century. These fall into two categories: books that distilled the information gathered by Ogilby's surveys into tables, and those that reproduced, with varying degrees of fidelity, Ogilby's strip-map format. The first category was dominated by multiple incarnations of *The Traveller's Pocket-Book; or Ogilby and Morgan's Book of the Roads*, first published in 1676 and re-issued, under various titles, up until the very end of the eighteenth century, reaching its 24th

incarnation in 1794. More relevant to the pictorial thread I am tracing is the second category, which made its debut in the year of *Britannia*'s publication with *Itinerarium Angliæ* – a condensed version of Ogilby that included only its maps, dispensing with its text and thus halving the weight. After a brief hiatus, this was followed by John Senex's *An Actual Survey of all the Principal Roads of England and Wales* of 1719, a simplified version of Ogilby's strip maps that went into multiple editions, the last appearing in 1775. In 1720, John Owen and Emanuel Bowen took Ogilby's maps as a pattern for their *Britannia Depicta, or Ogilby Improv'd*, a small quarto volume that used a simplified strip-map format but crammed additional information about local sights into each strip; this book went to at least 10 editions, with the last appearing in 1764. In 1767 Thomas Kitchen took up the baton with his re-publication of Senex's maps as *Kitchin's Post-Chaise Companion*, and, just as the century was drawing to a close, John Cary introduced a new twist to the genre with the publication of *Cary's Survey of the High Roads from London to Hampton Court*. Published in 1790, this book made visually explicit what had been implicit in Ogilby and his earlier followers by inserting lines of sight into the strips which indicated the precise point on the road from which prominent landmarks could be seen (Figure 2.4).

From even the most cursory glance at this array of publications it is evident that Ogilby's *Britannia* was the model, source, or inspiration for most of the route maps published in Britain during the eighteenth century. In fact, the strip map format was used far more frequently by British cartographers than by those of any other nationality, making it a particularly British mode of organizing the visual representation of motion through a landscape. For Ogilby's maps and their successors had a nationalistic aim as well. Originally intended to chart characteristic regional features and identify natural resources, as a collection they became a means to generate a picture of the nation as a whole. *Britannia*'s maps and their offspring encouraged and shaped recreational travel throughout the eighteenth century, and although

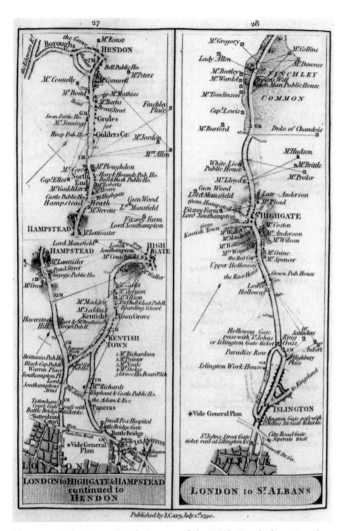

Figure 2.4 John Cary, Cary's Survey of the High Roads from London to Hampton Court *(London, 1790)*

Reproduced with permission of The British Library Board. Maps C.24.c.4.

I have not been able to ascertain exactly which wayfaring aids Gilpin may have used in his tours, it is unquestionable that by the time he embarked upon them Ogilby's '*Itinerary Way*', with its formulation of a journey as a succession of encounters with objects or sights along a route would have been a common mode of conceptualizing travel, and thus, by

extension, any motion through a landscape. Gilpin, however, configured the visual appreciation of this scenery in a new way. Gilpin's picturesque gave this mode of experiencing and understanding the British landscape a set of rules and standards, applying formal principles derived from painting to evaluate scenery viewed on the way from one place to another. His immensely popular books ignited the vogue for national tourism, and gave tourists a new object of pursuit: the British landscape itself. And in this way picturesque tourism gained a host of weightier consequences, for in the light of contemporary epistemological theories that stressed the correspondence between environment, behaviour, and identity, touring the national landscape became a means of getting to know not simply one's self as an individual, but, even more importantly, as a British citizen.

RIVER PANORAMAS AND CARTOGRAPHIES OF LEISURE

Gilpin's *Observations on the River Wye* established river travel as a popular touristic activity, defining its parameters, shaping expectations, and determining practices. Rivers like the Wye were paradigms of the picturesque, offering ever-changing scenery while being objects that were themselves in perpetual motion. But the picturesque was not an aesthetic applicable to all kinds of rivers, and its formulation of spatial experience as a series of individual moments threaded one after the other could not comprehend the blurred continuity of visual apprehension brought into being by new, mechanized forms of travel.[41] The publication of Gilpin's tours coincided with the development of the steam engine; by the time of his death in 1804 steamboats had begun plying the rivers and lakes of both Europe and North America, enabling groups of tourists to experience the aesthetic delights of river scenery *en masse*. Rivers conducive to steamboat travel were by necessity larger, wider, and straighter than the Wye and other rivers preferred by Gilpin, and as steamboat tourism gained in popularity, new kinds of cartographic modes were developed that reconfigured the representation of river travel for a new viewing public. In the early decades of the nineteenth century, Gilpin's picturesque fluvial aesthetics were combined with Ogilby's strip map format to create a new cartographic genre: the river panorama.

The 'Panorama of the River Thames from Windsor to the Nore', published as a fold-out section of *The Illustrated Hand-Book to London*, is a characteristic example of a type of river map that began to appear with increasing frequency in the 1820s (Figure 2.5).[42] Using a strip map format to represent a long stretch of the Thames, in this map the river's course serves as the line of the route. As with earlier strip maps, towns and objects of particular visual or historical note are depicted on both sides, but rather than being isolated points of picturesque interest, the vignettes are instead placed within the continuous landscape constituted by the river's banks. Eschewing the picturesque format of landmarks lined up in succession like beads on a string, the 'Panorama of the Thames' presents a convergence of landmark and landscape. The fact that this map, and others like it, were often referred to as panoramas is no coincidence. Not only did they begin to proliferate in the same decade that panoramas were enjoying their first great flush of popularity, but panoramic river maps aspired to convey the same kind of seamless and comprehensive overview offered by this other new form of representation. Like panoramas, panoramic river maps are continuous strips: printed on sheets that were glued one to another, the resulting long ribbon of paper was folded between boards like an accordion. And like panoramas, they are a modern phenomenon, one that has roots, as we have seen, in earlier cartographic forms, but that nonetheless constitutes a fundamentally new kind of viewing experience, a mode of seeing that, according to Stephen Oettermann, was to become the nineteenth century's dominant visual paradigm.[43]

Gilpin had been fully aware that larger waterways required a mode of perception different from that of the picturesque. 'In the immense rivers that traverse continents', he acknowledged, picturesque 'ideas are all lost':

> *As you sail up such a vast surface of water, as the Mississippi, for instance, the first striking observation is, that perspective views are entirely out of the question. If*

Figure 2.5 'Panorama of the Thames from Windsor to the Nore', from The Illustrated Hand-Book to London and Its Environs: With Fifty Engravings, and a Panorama of the River Thames from Windsor to the Nore *(London, 1853)*

you wish to examine either of its shores, you must desert the main channel; and, knowing that you are in a river, make to one side or the other.[44]

The visual experience offered by grand rivers such as these required a different orientation: passengers did not face up or downstream, in the direction of the boat's movement, but instead would have had to place themselves perpendicular to the route, facing one side of the river or the other. William Wade's *Panorama of the Hudson River* of 1846, published for the use of steamboat passengers on the popular tours from New York City to Albany, uses the formula of a strip map, but accommodates this change in viewing practice (Figure 2.6).[45] The river still serves as the line of the route, but rather than giving the banks a uniform orientation, Wade depicted them as two different strips, requiring passengers wishing to view the representation of the river's eastern banks to hold the map in one direction, while those wishing to examine the western side would have had to turn the map 180 degrees. Each bank of the river is rendered as a continuous elevation stretching from Albany to New York City, reflecting the fact that passengers could not stop the boat at will to tailor the trip to their liking, because they were now subordinated to the standardized structure of a

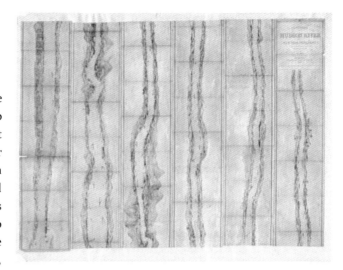

Figure 2.6 William Wade, Panorama of the Hudson River from New York to Albany *(New York, 1846)*

I.N. Phelps Stokes Collection, Miriam and Ira D. Wallach Division of Art, Prints and Photographs, The New York Public Library, Astor, Lenox and Tilden Foundations.

pre-established tour that moved with a uniform velocity and with equal ease whether the boat was travelling up- or downstream. Furthermore, although points of particular visual interest are still highlighted by the inclusion of

Figure 2.7 Wallace Bruce, Panorama of the Hudson Showing both Sides of the River from New York to Albany *(New York, 1910) (photo by Daniel Talesnik)*

Private collection.

identifying text, the distinction between landmark and landscape has become even less pronounced: instead, the landscape to either side of the river has been united into a horizontal continuity, opened up and flattened out like the pages of a book.

Whereas in 'The Panorama of the River Thames' we can detect the legacies of both Ogilby's strip map format and Gilpin's fluvial aesthetics, by the time of Wallace Bruce and G. Willard Shear's *Panorama of the Hudson from the Metropolis to the Capital* of 1888, modern technology has thoroughly altered the river panorama genre (Figure 2.7).[46] Harnessing the medium of photography for what claimed to be 'the first photo-panorama of any river ever published' this strip map standardizes the Hudson to a uniform width, and fuses points of picturesque interest with their surroundings. Although notable sights are identified by caption, and at times engravings of their elevations are collaged in, with the click of the camera's shutter mathematically synchronized to the speed of the steamboat, the resulting photographic strips give the impression of a two-dimensional continuity, of a representation that exemplifies Dolf Sternberger's characterization of panoramic perception, in which views 'have entirely lost their dimension of depth and have become mere particles of one and the same panoramic world that stretches all around and is, at each and every point, merely a painted surface'.[47] The steamboat and the camera have come together to generate a representation of landscape that approaches the cinematic, presenting, like the surface of a river reflecting its banks, two parallel, flattened, and seamless strips of scenery moving in time with the steamboat's engine, and transforming the experience of river travel into the passive spectatorship of a uniform and continuous visual flow.

ENDNOTES

1   William Gilpin, *Observations on Several Parts of England, particularly the Mountains and Lakes of Cumberland and Westmoreland, Relative Chiefly to Picturesque Beauty, Made in the Year 1772*, 3rd edition (London: Cadell & Davies, 1808), II, p. 100.

2   William Gilpin, 'A Fragment, Containing a Description of the Thames, between Windsor, and London: Accompanied with 37 Sketches, by Mr. Sawrey Gilpin', manuscript, National Art Library, Victoria and Albert Museum, 86 TT 18: i. In contrast to modern usage, Gilpin (along with many other seventeenth and eighteenth-century writers) habitually uses 'it's' with an apostrophe to signify the possessive.

3   Gilpin, 'A Fragment', p. 1.

4   'A Fragment', p. 3.

5   'A Fragment', p. 4.

6   'A Fragment', pp. 4–5.

7   William Gilpin. *Observations on Several Parts of England, Particularly the Mountains and Lakes of Cumberland and Westmoreland*, I, p. 210.

8   William Gilpin, *Observations on the River Wye, and Several Parts of South Wales, &c. Relative Chiefly to Picturesque Beauty; Made in the Summer of the Year 1770* (London, 1782). Although the date of publication that appears on the title page is 1782, because of problems with the aquatints the book did not actually appear until the following year. In addition to sketches found throughout the manuscripts of Gilpin's tours, largely held by the Bodleian Library, and the V&A's manuscript of the Thames tour discussed above, a bound volume devoted entirely to river scenery, containing 36 plates and seven pages of text, is held in the collection of the Morgan Library in New York (call number 1978.39).

9   Gilpin cites Polidore Vergil in 'A Fragment', I, and was clearly well aware of Camden's treatise. I am grateful to Michael Leslie, who long ago mentioned this aspect of Camden to me, and thereby planted the seed that eventually grew into this article. My thanks are also due to Andrew McRae, who drew my attention to William Harrison's 'Description of Britain'. For more on Harrison,

see Andrew McRae, *Literature and Domestic Travel in Early Modern England* (Cambridge: Cambridge University Press, 2009), p. 24.

10 *Journal of the House of Lords* (London, 1767–1773), XI, p. 675, quoted in T.S. Willan, *River Navigation in England, 1600–1750* (London: Frank Cass, 1964), p. 29, and discussed by Andrew McRae in *Literature and Domestic Travel*, p. 21.

11 The unrivalled source is still W.T. Jackman, *The Development of Transportation in Modern England* (London: Cass, 1916, 1962). For river travel, see Willan, *River Navigation in England*, and McRae, *Literature and Domestic Travel*.

12 Willan, *River Navigation in England*, p. 22.

13 Jackman, *The Development of Transportation in Modern England*, pp. 159–162.

14 W.G. Hoskins, *The Making of the English Landscape* (London: Penguin, 1955), pp. 247–254.

15 Jackman, *The Development of Transportation in Modern England*, pp. 171–176.

16 T.D. Fosbrooke, *The Wye Tour, or Gilpin on the Wye, with Historical and Archaeological Additions* (Ross, 1818).

17 See Malcolm Andrews, *The Search for the Picturesque: Landscape Aesthetics and Tourism in Britain 1760–1800* (Stanford: Stanford University Press, 1989), pp. 85–107.

18 Gilpin, *Observations on the River Wye*, p. 43.

19 Gilpin describes the work as 'the first of the kind I ever amused myself with', William Gilpin, *Observations on the River Wye*, ii, pp. 1–2. For more on Gilpin's tours, see Carl Paul Barbier, *William Gilpin: His Drawings, Teaching, and Theory of the Picturesque* (Oxford: The Clarendon Press, 1963), ch. 5.

20 Gilpin, *Observations on the Western Coasts of England*, p. 241; *Observations on the River Wye*, p. 8.

21 Gilpin, *Observations on Several Parts of England, Particularly the Mountains and Lakes of Cumberland and Westmoreland*, pp. 69–70.

22 Ibid., pp. 69–70.

23 Gilpin, *Observations on the River Wye*, p. 8.

24 William Hogarth, *The Analysis of Beauty* (London, 1753).

25 Thomas Whately, *Observations on Modern Gardening* (London, 1770), p. 72.

26 William Hogarth, *The Analysis of Beauty*, ed. Ronald Paulson (New Haven: Yale University Press, 1997), p. 42.

27 Ibid., p. 42; Home, Henry, Lord Kames, *Elements of Criticism* (Edinburgh, 1762), vol. I, p. 311.

28 Gilpin, *Observations on the Coasts of Hampshire, Sussex, and Kent, Relative Chiefly to Picturesque Beauty: Made in the Summer of the Year 1774* (London, 1804), p. 1.

29 Gilpin, *Observations on the River Wye*, p. 17.

30 In the eighteenth century Mount Cenis was the most frequented pass over the Alps on the way from France into Italy.

31 Gilpin, *Observations on the River Wye*, pp. 70–72.

32 Kames, *Elements of Criticism*, p. 21.

33 Gilpin, *Remarks on Forest Scenery*, II, p. 225, cited in Andrews, *The Search for the Picturesque*, p. 70.

34 The differences between the two terms are apparent from their etymology. Our common understanding of 'road' is a relatively modern variant, as road (*rad*, in Old Frisian, Middle Dutch *rede*, Middle Low German *ret*) originally meant not a thing at all but the act of riding on horseback – i.e. a road was a place where one rode. Route, instead, comes from the Latin *rupta*, meaning broken. A *via rupta* is a way opened by force. See 'road' and 'route' in the *Oxford English Dictionary* online.

35 The *Iter Britannicum* contained 15 itineraries measured out in Roman miles and paces (a Roman pace is two steps – left plus right – and a Roman mile is equivalent to 1,000 paces), documenting the routes between the province's major cities and military encampments. It was the only textual record of Roman infrastructure in Britain and, as such, was of seminal importance for later geographers, historians, and antiquaries.

36 John Ogilby, *Britannia, Volume the First: Or, an Illustration of the Kingdom of England and Dominion of Wales: By a Geographical and Historical Description of*

the *Principal Roads Thereof* (London, 1675). Ogilby and his team surveyed about two thirds of that, and published over 2,500 miles in all.

37 According to Ogilby the length of an English mile was computed using the module of a barley corn, of which '3 in length make an Inch, 12 Inches a Foot, 3 Feet a Yard, 3 Feet 9 Inches an Ell, 5 Feet a Pace, 6 Feet a Fathom, 5 Yards and a half or 16 Feet and a half a Pole, Perch, or Rod, 40 such Poles a Furlong, and 8 Furlongs a Mile; so that a Mile English contains 8 Furlongs, 320 Poles, 1056 Paces, 1,760 Yards, 5,280 Feet, and 63,360 inches' (and thus 190,080 barley corns) – a vivid example of the importance of agriculture to England's sense of itself as a nation. *Britannia*, p. 3.

38 For potential models and precursors, in particular the question of whether Ogilby knew Matthew Paris's itineraries, see Catherine Delano-Smith, 'Milieus of Mobility: Itineraries, Route Maps, and Road Maps', in James R. Akerman (ed.), *Cartographies of Travel and Navigation* (Chicago: University of Chicago Press, 2006), pp. 46–49, and Catherine Delano-Smith and Roger J.P. Klein, *English Maps: A History* (Toronto: University of Toronto Press, 1999), p. 170.

39 For a history of the strip map, see Alan M. MacEachren, 'A Linear View of the World: Strip Maps as a Unique Form of Cartographic Representation', *The American Cartographer*, vol. 13, no. 1 (1986), pp. 7–25, and Alan M. MacEachren and Gregory B. Johnson, 'The Evolution, Application and Implication of Strip Format Travel Maps', *The Cartographic Journal*, vol. 24 (December 1987), pp. 147–158. For a history of the itinerary map in the United States, see Nick Paumgarten, 'Getting There: The Science of Driving Directions', *The New Yorker*, 24 April 2006, available online at: www.newyorker.com/.

40 Although earlier commentators such as J.B. Harley and Katherine S. Van Eerde assumed that *Britannia* had been widely used *en route*, Catherine Delano-Smith and Garrett Sullivan have questioned this assumption. See J.B. Harley, 'Introduction' to the reprint of John Ogilby, *Britannia* (Amsterdam: Theatrum Orbis Terrarum, 1970), p. xvii; Catherine Delano-Smith, 'Milieus of Mobility', pp. 16–68; and Catherine Delano-Smith and Roger J.P. Klein, *English Maps: A History*, pp. 168–172.

41 Gilpin acknowledged that since England's rivers tended to be of the small and winding sort, extensive river prospects were rare: 'The continent of Europe may exhibit landscapes adorned with the Danube, the Rhine, and the Rhone - or rather landscapes, in which the water makes the principal part. But in England the land generally prevails'. William Gilpin, 'River-Views, Bays, and Sea Coasts', manuscript number 1978.39, The Morgan Library, New York. My thanks to Janike Kampevold Larsen for bringing this manuscript to my attention. For the changes brought to perception by the advent of the railroad, see Wolfgang Schivelbusch, *The Railway Journey: The Industrialization of Time and Space in the 19th Century* (Berkeley: University of California Press, 1986).

42 *The Illustrated Hand-Book to London and its Environs: With Fifty Engravings, and a Panorama of the River Thames from Windsor to the Nore* (London, 1853).

43 Stephen Oettermann, *The Panorama: History of a Mass Medium*, trans. Lucas Schneider (New York: Zone Books, 1997).

44 Gilpin, *Observations on the Western Parts of England*, pp. 238–239.

45 *Wade and Croome's Panorama of the Hudson River from New York to Albany. Drawn from Nature & Engraved by William Wade* (New York: Disturnell, 1846).

46 As a sequence of photographs bound as a book, it exhibits what Jonathan Crary has described, with respect to the Kaiserpanorama, as a 'structuring of experience common to many precinematic devices in the 1880s, and then to cinematic ones in the 1890s, in which the fragmentation of perception inherent to the apparatus is at the same time presented in terms of a mechanically produced continuum

that "naturalizes" the disjunctions'. *Suspensions of Perception: Attention, Spectacle, and Modern Culture* (Cambridge, Massachusetts: MIT Press, 2001), p. 138.

47 Dolf Sternberger, *Panorama, oder Ansichten vom 19. Jahrhundert*, 3rd edition (Hamburg: Claassen & Goverts, 1955), p. 57, quoted in Wolfgang Schivelbusch, *The Railway Journey*, p. 61.

3 'To Lead the Curious to Points of View': The Eighteenth-Century Design of Irish Roads, Routes and Landscapes

Finola O'Kane

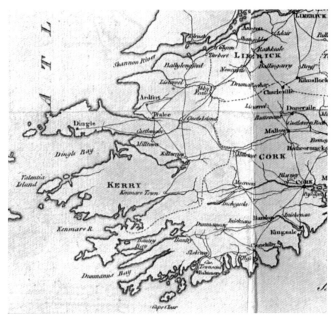

Figure 3.1 Taylor and Skinner, Maps of the Roads of Ireland, *2nd ed. (Dublin, 1783)*

Detail showing Kerry from the general 'Map of Ireland'. Courtesy of the National Library of Ireland.

Tourism in Ireland is a stage in the colonization of territory. An advanced stage, but part of the process by which a country becomes worth owning, worth investing in and finally worth presenting to outsiders. The Anglo-Irish ascendency found tourism to be a potent form of improvement, and a means of further civilizing the landscape and inhabitants of Ireland. What defined Ireland, and in particular justified her colonization, was her evident lack of application in improving herself. Charles Smith, writing from the English point of view in 1756, described 'the Irish nation' as 'universally acknowledged to be of great antiquity' with her inhabitants wanting 'neither wit, nor valour' and (unlike other heathen countries) having 'received the Christian faith as early as most countries of this western world'. The Irish had however never seen fit to 'inclose or improve their lands', and this situation required the improvement (and colonization) of Ireland by those who knew best how to go about it. Roads were a most effective form of improvement, uniting disparate parts of the poorly connected island, providing public works projects and altogether coordinating the use and representation of the landscape over which they passed.

THE ROAD TO KERRY

Kerry, located in the far south-west corner of Ireland, and described in 1756 as 'a terra incognita to the greater part of Europe'[1] was controlled by some seven families, most of them the recent beneficiaries of substantial Cromwellian land bequests (Figure 3.1). Several of these gentlemen 'since the spirit of improvement hath appeared in Ireland' proceeded enthusiastically 'in building, planting, inclosing, improving, and reclaiming waste and unprofitable ground, to the enriching of themselves, and adorning their country'.[2] They also began 'at great expence to carry on several very excellent roads through the country, without any public tax or other assistance, but by a voluntary subscription among themselves' allowing 'many inaccessible tracts' to be 'in some measure, restored to the kingdom, which were separated from it by impassable bogs and mountains'.[3] In the eighteenth-century Grand Juries, predecessors of the present-day county councils, were charged with laying out roads. With a membership comprising principally of the major landowners, the grand jury of each county was naturally desirous to connect their own properties to the road network as a matter of priority, and any concern for the public good lay contiguous to this private interest. The agricultural theorist and traveller Arthur Young found little to chastise in this system, finding it to result in what would have been desirable in any case:

> Every person is desirous of making the roads leading to his own house, and that private interest alone is considered in it, which I have heard objected to the measure; but this I must own appears to me the great merit of it [...] for a few years the good roads were all found leading from houses like rays from a center, with a surrounding space without any communication; but every year brought the remedy, until in a short time, those rays, pointing from so many centers, met; and then the communication was complete.[4]

Roads improved Ireland, not only because of connecting outlying areas to the kingdom, which might so far have escaped such stringent attachment, but also because the very act of making a road improved the adjacent countryside. Thus the principal road from Abbeyfeale to Killarney in Co. Kerry was applauded not only for being 'carried in direct lines, over mountains, through bogs, and morrasses' with 'several stone bridges', but also for being made 'with deep cuts, or ditches on either side, for the carrying off the water' which made 'the land on both sides [...] considerably drier than before'.[5] Roads, today, are not generally interpreted as constructions of aesthetic ambition. Yet, in a period such as the early eighteenth century, concepts of *in utile dolce* ensured that the useful was also beautiful, and a well-made and well designed roadscape not only connected towns for their material, geographical and political benefit, but was also a practical and therefore beautiful form of improvement. Road design was also considered an intrinsic part of the wider landscape design and a new road axis could become the initial design move for all that followed. Such a landscape may still be found in the flat plain of north Kerry where the east-west axis of Lixnaw town's main street, built as a dyke above the surrounding water-ridden landscape, became the deriving axis for a great constellation of roads, canals and avenues marching out from the central focus of the Earl of Fitzmaurice's Court of Lixnaw (Figure 3.2). For a very low-lying landscape such as that of Lixnaw, the fixed level of the road became the datum for the rest of the estate's design and in a terraced polder-inspired landscape, the landlord's demesne was reinforced by raising its level 'above the road', while other less significant areas were defined as 'lying below the road'.[6] This hierarchy of relative level affected the valuation of such landscapes, and their perceived success or failure as improvements depended upon such hierarchies being rigorously maintained.

The eminent Irish geographer John H. Andrews distinguished between 'evolved roads [...] that appear to have assumed their present course gradually, by a process of trial and error, as successive travellers have picked their way across the countryside' and 'the second kind of road' being 'the product of a preconceived design aimed at connecting two terminals in accordance with some definite principle, usually that of minimum distance, minimum gradient or a combination of both'. Andrews identifies a 'new "geometrical" style of road making that remained in favour at least until the 1760s' remarking that 'almost all the long straight roads of Ireland seem to belong to this period'.[7] Lixnaw's road, canal and avenue axes were laid out in the first half of the eighteenth century, and had reached some maturity by 1756 when described as situated 'agreeably on the River Brick', then 'cut into several pleasant canals, that adorn its plantations and gardens'. The improvements were 'very extensive' with 'most of the vistoes and avenues terminating by different buildings, seats, and farmhouses'.[8]

While Andrews does not conceive that any aesthetic design played some part in these decisions, finding that 'straight roads had the advantage of cheapness […] and of being easily laid out by crude surveying methods' he does observe that 'one notices that some of them […] seem to have been deliberately aligned, Roman fashion, on hill summits, and in one instance on a Norman motte'.[9] No Roman linked all the notable old and new monuments for many miles around into Lixnaw's legible diagram of Fitzmaurice power, influence and connectivity (Figure 3.2). At either end of 'the new road'[10] lay the Hermitage and the Monument, carefully flanking and framing the bulk of the demesne lands which lay south of the road, a prime orientation for an early eighteenth-century sequence of walled courts, gardens, orchards and summerhouses. Carefully positioned to benefit from the numerous prospects of mountains, ruined towerhouses and other monuments that lay to the south, Lixnaw's roads also defined the cross-axis of her great canal project, where the natural bed of the river Brick was usurped by two new lines of canal. These sidestepped carefully around the central house to allow the principal reception rooms long views over canals draining improved farmlands. Nor is it strange that they drove the straight routes through the bog, as the road was partly intended to ensure that the bog should no longer exist.[11]

A painting of Lixnaw, painted by Cornelius Varley in 1842 long after the Fitzmaurices had abandoned Kerry for Bowood in England, depicts the central function and force of the road itself in such a designed landscape (Figure 3.3). The 'new road' marching high above the original level of the surrounding countryside is shown to the right foreground, with two travellers gazing in the direction of the Old Court. The road of approach to this ancestral but now ruinous castle turns off to the left at a right angle further along our route, again at the same high dry datum. In the background, beyond the bridge over the canalized river Brick, stands the gaunt outline of a ruined house facade, on the site of the Hermitage. The water-logged fields below the road in all directions show that this intricate road/dyke landscape is not being maintained, the improvement has failed and the cattle themselves are taking up residence on the road. On the left hand side the ruined summerhouse and a bastion outcrop of the walled gardens together frame the distant mountains of Killarney. Bathed in nostalgia for a lost family and a lost landscape, the painting also reveals how such low-lying pseudo-Dutch landscapes demanded continued design conviction to retain their *raison d'etre*, and how initial ambition did not guarantee long-term practical success. Lixnaw's failure was due in part to the aesthetic ascendancy of the mountains Varley depicts so carefully in the distance, and to the energies of Killarney's resident Lord Kenmare when contrasted with the meagre interest of the then absentee Lord Fitzmaurice. Thus Killarney and not Lixnaw became the destination of choice for visitors to Ireland, and once removed permanently to England, the Fitzmaurice family, later Marquises of Lansdowne, could not direct infrastructural improvements as well as those living on the ground.

Penetrating outwards into the surrounding Kerry countryside, straight turnpike roads with tolled gates 'were laid out in 1748 from Millstreet via Castleisland to Listowel, with a branch to Killarney, while at about the same time other new roads, most of them equally straight, were cut by the local gentry from Castleisland to Abbeyfeale, Tralee and Killarney, and from Castlemaine to Tralee and Dingle' (Figure 3.1).[12] Lord Kenmare, when preoccupied with cutting 'a Canal from

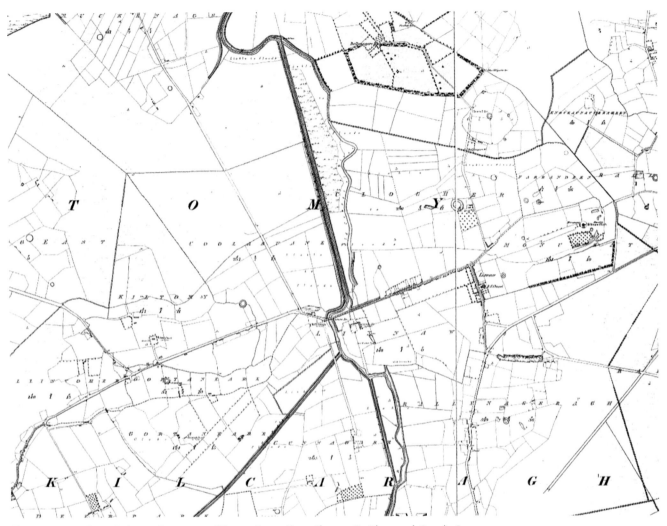

Figure 3.2 First edition Ordnance Survey map of Lixnaw, County Kerry, Sheets 15 & 16 (surveyed 1841–2), 1845
Reproduced courtesy of Trinity College Dublin.

his House to the Lake' of Killarney in order to 'confine the sight to the view of many objects of the beautiful scenery',[13] simultaneously set about building straight roads through his estates in the 1750s.[14] The success of Ireland's road-building programme was soon remarked upon by others, with Arthur Young observing that 'for a country so very far behind us as Ireland, to have got suddenly so much the start of us in the article of roads, is a spectacle that cannot fail to strike the English traveller exceedingly'.[15] Puzzled as to why such straight roads should have been 'tolerated and even admired for so long', the geographer Andrews remarks that it was only by the 1760s and 1770s that 'a new tendency was observable'[16] in favour of a more level rather than a straight route, which became naturally more curvilinear when the road swerved to

Figure 3.3 Cornelius Varley (1781–1873), Lixnaw Castle Co. Kerry, *oil on canvas (1842)*

Collection Ulster Museum. National Museums Northern Ireland 2011.

Figure 3.4 Two now disused Kerry roads, one following a straight line, the other adopting a more curvilinear approach (photo by Finola O'Kane)

Private collection.

the central axial focal-point to a moving shifting progression of contrivedly 'natural' scenes, so the roadscape changed to facilitate the new aesthetic.[17]

THE ROAD TO IRELAND

Unlike England, the understanding courted by Ireland's tourist literature is not that of the native but that of the well-heeled foreigner, and the perspective addressed, that of arrival by boat into Dublin harbour. Dublin, as the capital city, became the infrastructural centrepiece, and radial roads marched out from Dublin to the principal towns, and thence to the more minor hamlets. In 1777 George Taylor and Andrew Skinner were awarded a grant by the Irish House of Commons to survey and then produce 'Maps of the Roads of Ireland'.[18] The publication openly declared its loyalty to the protestant Anglo-Irish ascendancy with a frontispiece of the Boyne Obelisk, a memorial to the victory of England's protestant succession in the Glorious Revolution of 1688. The roads were ordered radially from the north to the south of Ireland and all roads led outwards from Dublin, with the north-point veering to the perspective of the outbound tourist (Figure 3.5). Carefully inscribed on each map were the seats, demesnes and names of the principal gentlemen, who both owned and organized the Irish landscape. One of the Kerry turnpike straight roads from the 1750s is clearly depicted in Taylor and Skinner's map of the 'Castleisland to Killarney' road, where the destination itself is awarded most of the geographical focus. Lord Kenmare's seat in the centre of the town, the other great houses, the town terraces, the roads leading out of Killarney and the great Lake of Killarney itself, become both the foci of the road and of the traveller's interest. The intermediary lands between departure and destination are almost blank, the reader and tourist encouraged to move rapidly from one to the other.

'As a cultural product, Maps of the Roads of Ireland [...] is self-consciously from "the gentleman's viewpoint"'. It is also

avoid hills. This change, from straight, and frequently steep, axial lines to curvilinear, level swoops, observed and dated by Andrews, mirrors precisely developments in landscape design and aesthetics. As the preferred point of view moved from an early example of the tourist's viewpoint, and the editing eye that ensured that the 'stretches between one landlord house and the next are apparently empty of people' with 'forest planting' as the 'only rural activity [...] represented'[19] is also that of the tourist, editing the landscape visually in his progress (Figure 3.3). In the eighteenth century, processional routes past key public buildings, churches, market houses and tholsels helped to define and express the protestant Anglo-Irish ascendancy identity in the urban context. The relatively new fashion for touring the countryside may have been inspired by that experience of the city. In the formal observance of tourism, such markers in the countryside were incorporated into books of views to reinforce that same identity, and to make the practice of tourism an activity guaranteed to reinforce the pattern of cultural hegemony in Ireland. Books of views projected the correct manner of interpreting the country to the outside world and ensured that its representation was framed in way that supported the existing ascendancy structure. The road infrastructure and its hierarchy of towns became 'the face of the country'[20] turning towards the foreign visitor.

Wilson's *The Post-Chaise Companion: or, Traveller's Directory, through Ireland* was then compiled from Taylor and Skinner's 'only authentic survey ever made of the roads of Ireland' in 1784. 'Similar to Mr. Paterson's useful Book of the English Roads' it expanded Taylor and Skinner's lists to include 'descriptions of Cities, Towns, Churches, Castles, Ruins, Gentlemens Seats, Loughs, Waterfalls, Spa's, Glens' and also set out to provide an account of the 'present State of Improvement, and Appearance of the Country, so essentially different from what it has hitherto been'.[21] The dual identity of writer as painter is dominant in the Preface: 'Our Picture, however, of the present State of the Country being just and faithful, with respect to the resemblance, however deficient in the colouring, we trust it will be as lasting as the nature of it will permit'.[22] As the introduction proceeds the literary imagination becomes yet more visual:

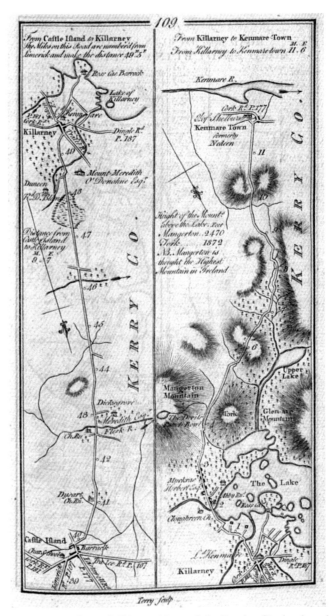

Figure 3.5 Taylor and Skinner, Maps of the Roads of Ireland, *2nd ed. (Dublin, 1783), 'From Castleisland to Killarney'*
Reproduced courtesy of the National Library of Ireland.

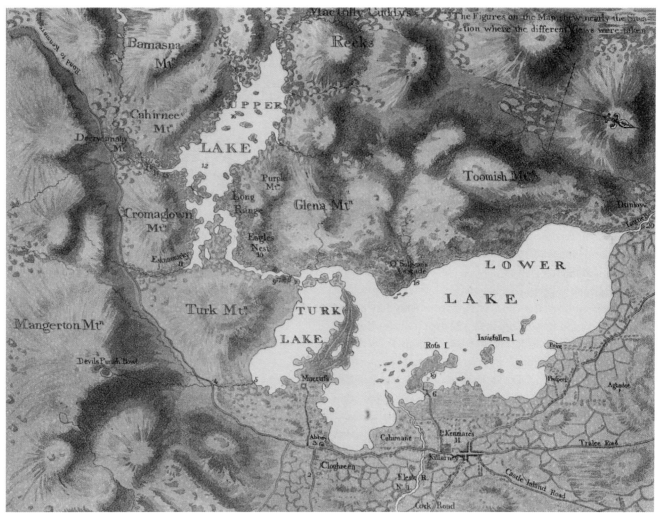

Figure 3.6 Map of Killarney

Jonathan Fisher, *A Picturesque Tour of Killarney describing in twenty views the most pleasing scenes of that celebrated Lake accompanied by Some general Observations and necessary Instructions for the use of those who may visit it; Together with A Map of the Lake and its Environs, Engraved in Aquatinta [...]* (London, 1789). Reproduced courtesy of the National Library of Ireland.

As new scenes, new circumstances arise, it may be re-touched and varied, according to the inclinations or abilities of the succeeding painters, the Contour, or Outline, being still preserved. The subject is susceptible of continual improvement; and particular parts of the work may receive it from persons that might be unequal to the whole performance. The ground-Work here laid may enable future artists, when the head and hand are cold that furnished these materials, to finish a picture that travellers may contemplate with pleasure.[23]

Wilson treats the road itself as a noteworthy object. Alternative routes are proposed, the longer one usually being the most picturesque, with 'the whole Road from Gilford to Warringstown' recommended for being 'ornamented with Rows of Trees on either Hand'.[24] Once the decision had been taken to structure the Ireland's routeways in such publications, the terminii of the routes acquired additional significance, in particular the terminus of all tourist routes to the south-east: Killarney. The potential of Wilson's envisaged link between text and image to both strengthen and derive the Irish tourist trails was first exploited by the painter and author Jonathan Fisher (d. 1809) A self-taught landscape painter, he published six large 'Views of Killarney' in 1770 and in 1789 *A Picturesque Tour of Killarney*, 'describing in twenty views the most pleasing scenes of that celebrated Lake [...] engraved in aquatint by Fisher himself'.[25] In two later editions the views were 'accompanied by Some general Observations and necessary Instructions for the use of those who may visit it Together with A Map of the Lake and its Environs' (Figure 3.6). Fisher did not 'propose to launch into description farther than is necessary to explain the Plates, but merely to lead the curious (who visit the Lake) to points of view, where the sublime and beautiful are most picturesquely combined'. Consciously encouraging the tourist to travel in his exact route Fisher also published a map carefully marked with the positions from which he had painted the views (Figure 3.6).[26] This map served to record and publish his points of view 'which often might be hastily passed by, if the Painter's observation did not induce a more critical examination'.[27]

Fisher acknowledged that the approach viewpoint 'on the northern line', for a visitor 'approaching the lake in this direction (which is also the approach depicted by Taylor and Skinner's map in Figure 3.5), after traversing dreary wastes, and arriving at a spot where the first appearance of a long-wished for object presents itself, the eye naturally feels relieved, and often conceives a partiality for the first interview [viewpoint]'. He then further refined the visitor's experience by advocat-

Figure 3.7 Jonathan Fisher, 'No. 1 North View of the Lake of Killarney' from Appendix to Scenery of Ireland *(Dublin, 1796), from Jonathan Fisher,* A Picturesque Tour of Killarney *(1789)*

Reproduced courtesy of the National Library of Ireland.

ing a viewpoint limited to that lying on the approach road from the north between the outside limits of Lord Kenmare's park to the east and the ruins of Aghadoe to the west, 'beyond which the open between Glena and Turk [mountains] is lost to the eye, and the expanse to the distant mountains precluded from view'.[28] Fisher painted the view from this exact position on the road for his publication *Scenery of Ireland*, entitling the view 'No. 1 North View of the Lower Lake of Killarney' (Figure 3.7). In the hierarchy of landscape beauty his publications helped to bring about, this view was the leading position from which his choreography of tourism began to dance.

TOWNSCAPES ON THE ROAD

Roads continued to derive the view into the nineteenth century. Glin, a town situated on the 1764 'mail road, leading from Limerick to Tralee',[29] and an important secondary route to Killarney, worked hard to improve its road profile. Taylor and Skinner's road map of 1777 marks both the dotted line of

'The Old Road' and the line of the new road which runs along the shore of the Shannon estuary 'enclosed off the strand and raised about four feet above the level of high water [tide]'.[30] Depicted with pride in the commissioned estate painting of 1839 by Jeremiah Hodges Mulcahy (Figure 3.8), the route of this new road snakes across the painting from walled parkland to dark woodland and emerges onto the brilliant causeway of Glin's new sea wall, road and embankment. The estate itself, with its white great house appearing above the new demesne sea wall, also responded to the new road. There the demesne wall became a retaining wall, as the road runs below the level of the interior parkland, and substantial earthmoving effected the contrivance of the road's landscaped section. By casting the road as a haha to ensure that it could not impinge on any views northwards from the house and its parkland, the historic connection between estate and waterfront diminished, and the road effectually formalized the demesne's relationship with the estuary. Other external estate buildings also felt the pull of the road, and the eye-catcher of Hamilton's tower (1822) draws the gaze not only of those within the estate but also those moving along the road (Figure 3.9). The sea road's construction may also have curtailed the inhabitants' enjoyment of the estuary, and none of the bathing or boating houses so common to nineteenth-century waterside demesnes were built after its construction. However, the sea was still seen as a mixed blessing in the early nineteenth century. While the French traveller De Latocnaye observed that 'the neighbourhood of the flat and sandy bay of Ballyheigh' in Kerry had brought some sea-bathing visitors to Kerry in 1796,[31] a local resident thought that her brother would 'shudder' at 'the Idea' that she had 'Got a House close to the Sea' in 1805.[32]

In 1756 Charles Smith classified the towns of Kerry into 'large towns, seaports, and great-road towns',[33] suggesting that a town's only distinction might be a road running through it. Being on the road was a considerable advantage, and landlords made considerable efforts to control the land between their estate and (ideally) the Dublin road. Without

Figure 3.8 Detail from Jeremiah Hodges Mulcahy, RHA, 1804–1889, A View of Glin and Shannon Estuary, oil on canvas, 1839
Reproduced courtesy of Desmond FitzGerald, The Knight of Glin.

Figure 3.9 Photograph of Hamilton's tower taken standing within the demesne looking east (photo by Finola O'Kane)
Private collection.

the controlling interest of a resident landlord towns could become unhitched from the developing road network, and a town's status, not infrequently, reflected that of its landlord, made further legible by the evolving approach routes, their controlling vistas and the published books of views, road maps and tourist literature. Without a resident, confident and powerful landlord, towns lost status, losing their markets and their ability to negotiate for improved infrastructure and their position on the road network. In Ireland 'roads approaching a demesne are suspiciously numerous' and 'when large private parks became fashionable, a gentleman living near the main highway might cause it to be diverted to a more fitting distance from his house'.[34] Some vague presentiment that other factors may be influencing the design of seemingly practical roads is evident in such observations that in the 'rejuvenated topography of Ireland, approaches to river crossings presented the most difficult problems, especially where the valley sides were occupied by demesnes, which the new roads were not allowed to penetrate'.[35] This predilection among eighteenth-century road designers for avoiding the landlord's demesne and its approaches, routes and framed views and thereby inducing illogical road redirections and large sweeping diversions, continues to puzzle those who prefer art and science to remain strictly segregated. Most of the confusion lies with the insistent interpretation of a road as a wholly practical project for moving goods and people from one place to another as expediently as possible. Yet 'we travel for various purposes - to explore the culture of soils, to view the curiosities of art, to survey the beauties of nature, and to learn the manners of men, their politics and modes of life'.[36] Ireland's river valleys were the locus of concentrated achievements of landscape design, and therefore of interest to her burgeoning tourist industry; road design in these areas had to be much more calculated to 'lead the curious to points of view' than to serve merely an anticipated traffic load and volume. Where considerable design energy had been expended on demesne landscape design, then naturally enough, the road design formed part of that larger project, and by being funded primarily through the private interest ensured that the client considered it part of his or her overall design activities. Thus Irish eighteenth-century road and route design had to take cognizance of landscape aesthetics to ensure that the landscape could be admired from the designed distance and viewpoint, causing the roads to lead

to it and not through it. The cast of characters involved in Ireland's eighteenth-century road design is also suspiciously design-proficient. Andrews remarks on the achievements of Charles Vallencey (1725–1812), who recorded a new curvilinear road being laid out 'to avoid the hills obstructing the old road' in his Military Survey. Yet military engineering was not Vallancey's only professional persona, he was also a landscape designer capable of advising Viscount Fitzwilliam on a 'little Improvement of the Pallisade, instead of the Wall, between the House & Stables of Merrion', which added 'such a Beauty at the Entrance and from the Road, up the Avenue'.[37] A member of the army corps of engineers who had published books on inland navigation and advised on many infrastructural projects,[38] such men did not distinguish precisely between practical and aesthetic design motives. It is unlikely that someone with the design sensibility of Vallencey, so careful of the design of the approach to Lord Fitzwilliam's mansion of Mount Merrion, would not have appreciated a road's influence over the interpretation of landscape by the eye of the tourist.

THE PICTURESQUE ROADSCAPE

Roads were picturesque in themselves, with William Gilpin encouraging its enthusiasts to 'break the edges of the walk: give it the rudeness of a road; mark it with wheel-tracks; and scatter around a few stones, and brushwood', summarizing that 'in a word, instead of making the whole smooth, make it rough; and you make it also picturesque'.[39] In his description of the approach to Hackfall from Studley in Yorkshire, Gilpin also explains the road as scene-setter for what follows:

It is a circumstance of great advantage, to be carried to this grand exhibition (as you always should be) through the close lanes of the Rippon road. You have not the least intimation of a design upon you; nor any suggestion, that you are on high grounds; til the folding-doors of the building at Mowbray-point being thrown open, you are struck with one of the grandest, and most beautiful bursts of country, that the imagination can form.[40]

As Gilpin suggests, roads also set the scene for what followed. Once tourism became more general, and capable of being promoted by books and images, the stage was set for a well-tempered roadscape. Pleasurable anticipation became a cornerstone of the tourist's experience, as did the designed, proscribed and wholly anticipated view. Once the Kerry landlords began to see tourism as an improvement the roads could not but respond to their viewpoint. Lord Fitzmaurice began by connecting his road design to that of his landscape and Lord Kenmare made straight axes of his turnpike roads because that was also his habit with his avenues, prospects and canals. The aesthetic of straight lines of hedges, trees and roads moving over both undulating ground and flat plain was improvement writ large upon the landscape. As the route itself, the road rarely escapes comment. Writing on vistas of Irish ruins the 1763 tourist Salusbury Brereton remarked that 'either Abbys or other Churches' and 'innumerable' castles were 'constantly to be found in compleat Ruin every 3 or 4 miles you ride all through' Ireland 'prov[ing] very amusing objects to travellers who otherways would be heartily tired of Dreary flat Views for 20 miles together, Tho on such excellent Roads'.[41] Criticising the Wicklow tour in 1823, the Rev. George Wright lamented the loss of the old hillside paths 'cut, in an irregular and picturesque manner, and the wild rustic road which ran from end to end', which had been 'extremely beautiful and appropriate'. These had been 'sadly altered by the present noble proprietor, Lord Powerscourt, in order to permit his Majesty's carriage to drive through, which could not have been done with safety along the old road'. This 'great sacrifice of part of the romantic beauties of the Dargle' had not 'afford[ed] the expected gratification, as time would not permit his Majesty to visit the delightful scene'.[42]

In 1779 Arthur Young wrote of his personal way of using maps and the roads and landscapes they represented:

I found it perfectly practicable to travel upon wheels by a map. I will go here; I will go there; I could trace a route upon paper as wild as fancy could dictate, and everywhere I found beautiful roads without break or hindrance, to enable me to realise my design.[43]

Young reveals here something of the essence of eighteenth-century travel – it was in itself a design project. What road you chose, how you connected to other routes, how you travelled along it, how you gazed from it, and what views you chose to pass by were design decisions, conscious choices of direction, orientation, view and experience. Being on the road was a desirable prospect for both tourist and landowner, as the design of roadside views served to alert visitors to the attraction of the area, and the achievement of the landowner in making it so. Thus eighteenth-century Irish houses, towns and landscapes start to turn self-consciously towards the road, and the published travel literature and books of views ensured that the tourist's gaze was restricted to a carefully designed programme of visual effects. This well-tempered roadscape helped to establish Ireland as a premier tourist destination for nineteenth-century Britain, where the road itself becomes a key piece of aesthetic infrastructure, using form, profile and orientation to derive the design events that attend upon its route.

ENDNOTES

1 Charles Smith, *The Antient and Present State of the County of Kerry. Being a Natural, Civil, Ecclesiastical, Historical and Topographical Description thereof* (Dublin, 1756), p. iv.

2 Smith, *The Antient and Present State* (1774 edition), pp. 66–67.

3 Ibid., p. 65.

4   Arthur Young, *A Tour in Ireland 1776–1779*, vol 2/2 (facsimile of 4th edition [London: Bell, 1892], Shannon: Irish University Press, 1970), p. 79.

5   Smith, *The Antient and Present State of the County of Kerry* (1756 edition), p. 169.

6   National Library of Ireland (NLI), p. 8575: *Maps of Lixnaw, Co. Kerry and Adjacent Lands Surveyed for Dudley Ryves by Charles Frizell Senior and Richard Frizell*, 1763.

7   John H. Andrews, 'Road Planning in Ireland before the Railway Age', *Irish Geography* v/1 (1964), pp. 17–41.

8   Smith, *The Antient and Present State* (1979 edition), p. 112.

9   Andrews, 'Road Planning in Ireland'.

10 NLI, p. 8575: *Maps of Lixnaw, Co. Kerry and Adjacent Lands Surveyed for Dudley Ryves by Charles Frizell Senior and Richard Frizell*, 1763.

11 Andrews, 'Road Planning in Ireland', p. 30: 'Apart from their straightness, the most strikingly distinctive feature of these early planned highways was their indifference to bog'. See also D. Broderick, *The First Toll-Roads: Ireland's Turnpike Roads, 1729–1858* (Cork: Collins Press, 2002).

12 Andrews, 'Road Planning in Ireland', p. 32.

13 Barton, Richard, *Some Remarks, Towards a Full Description of Upper and Lower Lough Lene, Near Killarney, in the County of Kerry* (Dublin, 1751), p. 14.

14 *Kenmare Mss.* ed. Edward Mc Lysaght (Dublin: Irish Manuscripts Commission, 1942), pp. 201, 205.

15 Arthur Young, *A Tour in Ireland*, p. 77: 'Section IX. Roads-Cars'.

16 Andrews, 'Road Planning in Ireland', p. 32.

17 For a detailed discussion of eighteenth-century Irish landscape design see: Finola O'Kane, *Landscape Design in Eighteenth-Century Ireland; Mixing Foreign Trees with the Natives* (Cork: Cork University Press, 2004).

18 Taylor and Skinner, *Maps of the Roads of Ireland*.

19 Jacinta Prunty, *Maps and Map-Making in Local History* (Dublin: Four Courts, 2004), p. 70.

20 William Gilpin, *Observations on the River Wye and Several parts of South Wales, &c. Relative Chiefly to Picturesque Beauty; Made in the Summer of the Year 1770* (5th edition, London, 1800), p. 1: 'a new object of pursuit; that of examining the face of a country by the rules of picturesque beauty: opening up the sources of pleasure which are derived from the comparison'.

21 William Wilson, *The Post-Chaise Companion: or, Traveller's Directory, through Ireland* (London, 1784), p. v.

22 Ibid., p. vi.

23 Wilson, *The Post-Chaise Companion*, pp. vi–vii.

24 Ibid., p. 18.

25 Walter George Strickland, *A Dictionary of Irish Artists* (London, 1913), vol. 1/2, p. 344: 'Two later editions of the work were issued, one with a map and with slight variations in the wording on the title page, published by Jones in Dublin in 1790 [...] and another published in London in 1791'.

26 Jonathan Fisher, *A Picturesque Tour of Killarney describing in twenty views the most pleasing scenes of that celebrated Lake accompanied by Some general Observations and necessary Instructions for the use of those who may visit it; Together with A Map of the Lake and its Environs, Engraved in Aquatinta [...]* (London: G.G.J. and J. Robinson, no. 25 Paternoster Row, 1789).

27 Jonathan Fisher, *A Picturesque Tour of Killarney* (London 1789), Introduction.

28 Jonathan Fisher, *Scenery of Ireland* (London, 1795): from Appendix description of *The Celebrated Lake of Killarney*, pp. 11–12.

29 Tom Donovan, 'The Evolvement of Glin Village' in *The Knights of Glin; Seven centuries of Change*, ed. Tom Donovan (Glin: Glin Historical Society, 2009), p. 353.

30 Tom Donovan, 'The Evolvement of Glin', p. 353: from Michael Mc Quane, 'Tour of Ireland by John Harden in 1797', *Journal of the Cork Historical and Archaeological Society*, Part II, vol. LIX, no. 190, July–December 1954, p. 69.

31 De Latocnaye, *A Frenchman's Walk through Ireland 1796–7*, translated from the French of de Latocnaye by John Stevenson (Dublin: Hodges, Figgis & co, 1917), p. 112.

32 Arabella Ward to [her brother] John Crosbie, Brighton, 14 January 1805, NLI , Talbot-Crosbie Mss. PC 188, f. 135.

33 Smith, *The Antient and Present State* (1774 edition), p. 78.

34 Andrews, 'Road Planning in Ireland', pp. 21–22.

35 Ibid., p. 35.

36 Gilpin, *Observations on the River Wye*, p. 1.

37 Charles Vallancey to Lord Fitzwilliam, 18 November 1764 (NAI, Pembroke Estate Mss., Ms. 97/46/1/2/6/70).

38 *Dictionary of Irish Biography*, 2009, Charles Vallencey entry.

39 William Gilpin, *Essays on Picturesque Beauty; &c. &c. &c.: Three Essays: On Picturesque Beauty; on Picturesque Travel; and on Sketching Landscape: To which is added a poem, on Landscape Painting, Second Edition By William Gilpin, A.M. Prebendary of Salisbury; and Vicar of Boldre in New Forest, near Lymington* (London: Printed for R. Blamire, in the Strand, 1794), Essay 1, p. 7.

40 William Gilpin, *Observations, Relative Chiefly to Picturesque Beauty, Made in the Year 1772*, Sect XXVI.

41 'Owen Salusbury Brereton Esqr. F.R. & A.S. late Vice President of the Society for the Encouragement of Arts Manufactures & Commerce; The Irish Tour of an Eighteenth-Century Antiquary and His Notions about Irish Round Towers, 1763' in The Knight of Glin, editor, *Ireland of the Welcomes* 19/3 September-October (1970), p. 11.

42 Rev. George N. Wright, *Tours in Ireland or, Guides to the lakes of Killarney; the County of Wicklow; and the Giants Causeway, illustrated by maps; and engravings, after drawings by George Petrie* (London, 1823): Guide to the County of Wicklow, p. 14.

43 Young, *A Tour in Ireland*, p. 79. 'For and expansion of many of the themes and sites of this paper please see Finola O'Kane, *Ireland and the Picturesque; Design, Landscape Painting and Tourism in Ireland 1710–1830*, Yale University Press on behalf of the Paul Mellon Centre for Studies in British Art (forthcoming).

4   A Road with a View: C.F. Vogt's Painting of Krokkleiva

Torild Gjesvik

Roads abound in landscape paintings. Once you start looking for them, they seem to be everywhere. Roads represent, among other things, a conventionalized way of bringing the viewer into the picture and leading him through the landscape. This is also true of the road featuring in the Danish artist Carl Frederik Vogt's picture *Krokkleiva*, painted around 1812–1817 (Figure 4.1). My interest in this painting, however, is much more specific. Vogt's picture does not just represent any road, in any landscape. It portrays a particular road, Krokkleiva in Norway, and was painted for the man who built it.

The merchant and politician Peder Anker (1749–1824) was one of the most prominent men of his time. He belonged to one of Norway's leading families, owned large estates and ran a flourishing business centred on timber trade and iron works. When the Union with Sweden was established in 1814, Anker was appointed Prime Minister of Norway and stationed in Stockholm.[1] In addition to his private business and his work as a politician, he showed a lifelong commitment to the improvement of the Norwegian road system. Vogt's *Krokkleiva* was part of a larger commission of paintings for the representative dining room at Bogstad Gård, Anker's estate and residence on the outskirts of Christiania, present-day Oslo.[2] Vogt painted eight pictures for this room, all still in their original places.

The context in which Vogt's painting was made and used makes it particularly well suited for investigating the relationships between the practice of painting roads and landscapes, the practice of road-building,

Figure 4.1 C.F. Vogt, Krokkleiva, *100 x 187.5 cm, oil on canvas, ca. 1812–1817, Bogstad Gård, Oslo (photo by Arthur Sand)*

Reproduced with permission of Bogstad Gård.

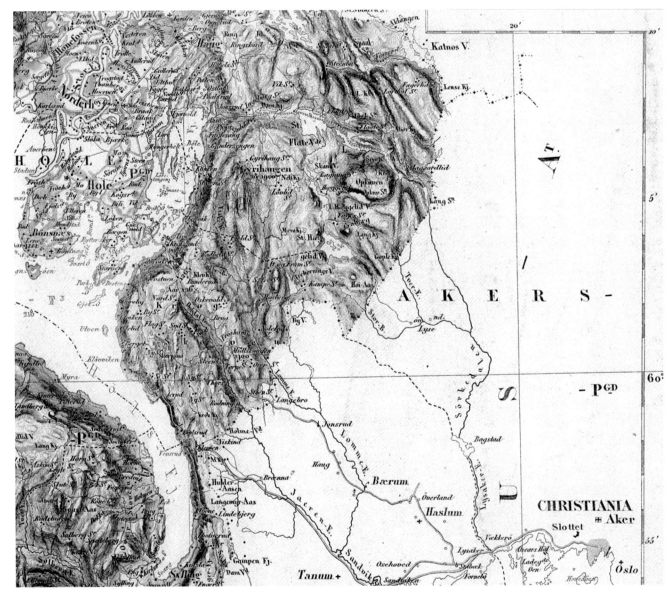

Figure 4.2 S.C. Gjessing, map of Buskerud Amt (detail), 1854

The road from Christiania to Ringerike appears to the left of Christiania. It passes Bærum and several small farms before it reaches the steep descent Krokkleiva leading down to the inlet Kroksund and the lake Steinsfjorden. Reproduced with permission of the Norwegian Mapping Authority, Hønefoss.

and the practice of looking at landscapes and roads. My aim is to suggest how these different practices are interrelated. I will do that by considering Vogt's painting of Krokkleiva in relation to its physical context in the dining room at Bogstad, to Anker's practice as road-builder and to the Anker-family's use of the road. The painting brings practical and aesthetic interests in the road and the landscape together in a very striking way. By highlighting these concerns, the paper will also attempt to elucidate the role of infrastructure in the aestheticization of the modern landscape.

KROKKLEIVA IN SITU

Before looking closely at Vogt's painting, a short presentation of the road itself is called for. Krokkleiva is an infamously steep stretch of the old route leading from the Norwegian capital, through the forest area Krokskogen, and down to the lowlands at Ringerike (Figure 4.2). The road featuring in Vogt's painting was finished by Anker around 1805. During the nineteenth century Krokkleiva became renowned throughout Europe for its beautiful views and its dramatic descent. It attracted painters, poets and tourists alike and has been termed the first Norwegian tourist road.[3]

Vogt's painting is among the earliest known depictions of Krokkleiva, representing an early phase in Norwegian landscape painting.[4] It shows us the road as a curve in the foreground. Dark, precipitous cliffs border the road, leaving it in deep shadow. Having passed through this narrow, gate-like passage, the road disappears as it descends steeply, opening up the view to the landscape below. A small part of the road can be seen a bit further ahead, this time in sunlight. It seems to be running at the edge of a cliff before it vanishes once again. The cliffs, framing both the road and the view, create a dramatic contrast with the sunlit lowland of Ringerike, the lake Steinsfjorden, and the bright sky above. Far off in the horizon we can see the mountains of Norefjell. Although the painting is heavily darkened by old varnish and dirt, we

Figure 4.3 The dining room at Bogstad Gård (photo by Arthur Sand) From left: Vogt's paintings of Bærum's Verk, Anker's hunting lodge at Stubdal, Krokkleiva and Vækerø. Reproduced with permission of Bogstad Gård.

can discern several figures in the foreground: A man walking alongside his horse, a woman on horse-back, and a couple of other figures sitting or standing by the road-side. The tallness of the format and the smallness of the figures further enhance the drama of the landscape, with its towering cliffs.

In the dining room at Bogstad *Krokkleiva* appears on the north wall, to the right of a semicircular niche. Its tall portrait format is echoed in the painting on the other side of the niche – an image of Anker's hunting lodge at Stubdal in Åsa, not far from Krokkleiva (Figure 4.3). These two, as well as the six other paintings by Vogt, are arranged symmetrically and have been made to fit the wall space. The main theme of the series is the presentation of estates and establishments belonging to Anker. They are shown as part of the surrounding landscape or as settings for feasts and social gatherings. The four large panels flanking the double enfilade doors show us two versions of Bærum's Verk, the ironworks owned by Anker (Figures 4.5 and 4.8); a view of Vækerø, the estate used by Anker as a harbour for his timber trade (Figure 4.4); and a view of Jarlsberg, the country estate owned by Anker's son-in-law, count Herman Wedel-Jarlsberg.[5] By the windows, overlooking the garden, there are two narrow panels, both showing imaginary and symbolic landscapes (Figure 4.8).

One of them is featuring two trees, one living and one dead, the other an obelisk in a cave.[6]

Although the paintings in the dining room at Bogstad are considered Vogt's major work, little has been written about them, and hardly anything about the Krokkleiva-painting.[7] Practically nothing is known about the commission and the relationship between Vogt and Anker. No written contract is known to have existed, and only two of the paintings are signed and dated.[8] However, judging by the paintings themselves, their setting and the artistic tradition they are a part of, they served at least three different functions: as ornaments, as a form of self-representation and as conversation-pieces. Anker was renowned for his hospitality and the dining room was frequently used for representative purposes. As well as being the administrative centre of Anker's extensive activities, Bogstad served as one of the most important meeting places for the Christiania elite, regularly receiving his business partners, fellow politicians and foreign guests, for example Mary Wollstonecraft and Thomas Robert Malthus.[9] The Swedish Crown Prince, Carl Johan (1763–1844), dined at Bogstad shortly after the Union with Sweden had been proclaimed and Anker had been appointed Prime Minister of

Norway.[10] Although the Crown Prince and his retinue would most likely have dined in the hall, and not in the dining room, it is not unlikely that he was shown Vogt's paintings, some of which would have been completed at this time.[11]

Having established the Krokkleiva painting within the frame of the dining room, it is worth asking how it relates to the other pictures in some more detail. As it is not my intention here to give a complete analysis of the series, or indeed the whole room, I will primarily attempt to view Vogt's paintings in a road perspective. Letting the image of Krokkleiva lead the way, so to speak, in order to see how this might enrich our understanding.[12]

ROUTES, ROADS AND LANDSCAPES IN THE DINING ROOM AT BOGSTAD

In addition to the Krokkleiva picture, two paintings in the dining room are of special interest to the student of routes and landscapes: The image of Vækerø (Figure 4.4) and of Bærum's Verk (Figure 4.5). Roads are central motifs in both of these. They do not appear as a mere result of compositional

Figure 4.4 C.F. Vogt, Vækerø *(detail), 191 x 187.5 cm, oil on canvas, ca. 1812–1817, Bogstad Gård, Oslo (photo by Arthur Sand)*
Reproduced with permission of Bogstad Gård.

Figure 4.5 C.F. Vogt, Bærums Verk *(detail), 191 x 187.5 cm, oil on canvas, 1813, Bogstad Gård, Oslo (photo by Arthur Sand)*
Reproduced with permission of Bogstad Gård.

strategies, but have something important to tell us about Anker's interests and activities. In *Vækerø* Vogt shows us the estate and the surrounding landscape in a sweeping panorama. The painter-beholder seems to be standing on a hill, looking down towards the buildings and the harbour at Vækerø, situated in the middle ground. Beyond the harbour we can see the peninsula Ladegaardsøen (present-day Bygdøy), and the Christiania-fjord.

A road appears at the left, drawing a curve before continuing straight on towards the harbour. It crosses another road (the main road between Christiania and Drammen), then runs down to the houses. At first sight the roads in this painting do not seem to beg particular attention. However, knowing that the road leading down to the harbour is the road that Anker had built shortly after he bought Bogstad, this pictorial element takes on a new meaning.[13] This was the first road Anker had built, and it gained him much praise at the time. It was probably decisive for his later appointment as *Generalveiintendant* [Intendant General of the roads in Norway], an office and a title created particularly for Anker in order to show his superior status.[14] The road was a vital link between two of Anker's estates. In the painting it is also shown as a part of his wider economic enterprise: The harbour with its tall piles of planks, the ships in the harbour and out on the fiord – on their way to Anker's trading partners abroad. In the painting of Vækerø, issues of landownership, economy and communication are embedded in the landscape, and the road running through it is important in linking these concerns together. Going back to the road itself, it is interesting to note that in contemporary texts it was described in such terms as 'a beautiful, string-like road'.[15] This sort of vocabulary shows that the road was considered not only from a utilitarian point of view but also from an aesthetic.

In the painting of Bærum's Verk (Figure 4.5) the road dominates the foreground of the picture. An open carriage – a phaeton – pulled by two horses, is approaching the ironworks, the buildings of which we can see in the middle ground. The two men sitting in the carriage has been identified as Peder Anker himself (dressed in black, and – significantly – holding the reins) and his administrator Niels Lasson (1762–1853).[16] Lasson is pointing in the direction of the ironworks, thus directing our attention towards it. The road depicted in this painting is also of Anker's making. It is Ankerveien – named after Anker himself – leading from Bærum's Verk, via his estate at Bogstad, to Maridalen, a valley just north of Christiania.[17] Ankerveien was also instrumental in linking together different parts of Anker's estates and was used particularly for the transportation of iron from Bærum's Verk to the rod iron hammer he had constructed in Maridalen.

The paintings of Vækerø and Bærum's Verk point to Anker as a private road builder, whereas the road depicted in the Krokkleiva-painting was public. It is worth noting however that the estates of Anker extended as far as Ringerike and even further. Krokkleiva was thus within Anker's zone of influence and he used the road as well as the nearby landscape in his private business activities. The most conspicuous example of this is the unique technological device called *Kjerraten* – a log haul up – that he erected at Åsa, in order to transport timber from his forest estates.[18] He also practiced hunting at his hunting lodge in the same area (Figure 4.3). The painting of Krokkleiva thus relates to Anker's wider activity in the region. By including it in his dining room he appropriated the road in a representational and symbolic sense.

PLACING THE ROAD IN THE LANDSCAPE: ANKER AS ROAD DESIGNER

I will leave the dining room for a moment, and look at certain aspects of Anker's work with Krokkleiva in relation to how the road is represented in the painting. First a few words about Anker's public role as a road builder. In 1789 he was appointed *Generalveiintendant* and until 1814 he held the highest authority on all matters concerning road building in the greater part of southern Norway.[19] His mission was

Figure 4.6 Wilhelm Maximilian Carpelan, Krogskoven, Vu de Ringerige, *21 x 29 cm, coloured aquatint, 1821–23, from* Voyage pittoresque aux alpes norvégiennes *(Stockholm 1821–23)*
The National Library of Norway, the Picture Collection. Reproduced with permission of The National Library of Norway, Oslo.

to build new roads and to improve old ones in his area. A particular emphasis was put on the improvement of the main roads – the so called *Kongeveier* [King's Roads]. The route from Christiania to Bergen (the main route between the eastern and the western part of the country) was one of the most important of these, and Anker was responsible for upgrading the larger part of it from a riding path to a carriage road. The new road down Krokkleiva – represented in Vogt's painting – was a part of this work.[20]

Anker's Krokkleiva is an example of the so-called French principle practiced in Norwegian road building from about 1760 to 1850.[21] According to this principle roads should be constructed as straight lines – cutting through the landscape – more or less independent of the steepness of the terrain. Anker adhered strictly to the French principle, and Krokkleiva is an extremely good example. W.M. Carpelan's print *Krogskoven, seen from Ringeriget* illustrates this (Figure 4.6). It is worth noting that the literature on

Anker agrees that he did not only use his position as *Generalveiintendant* to administer the work, but took a very active part in the designing of the roads. He made decisions about the trajectory of the roads and about construction methods; and he travelled extensively in the area appointed to him, in order to gain knowledge of the local conditions.[22] Anker also introduced new requirements in Norwegian road building: The roads were made broader, they received a gravel surface, and ditches were made at both sides of the road to ensure good drainage.

Some of these features can be seen in Vogt's representation of Krokkleiva. The road is quite broad, the road surface appears to be of a relatively good quality and we can see a ditch on the right hand side. However, the straightness of the road is not reflected in Vogt's painting. Instead the road curves as it enters the painting. The extreme steepness of the road – so obvious in Carpelan's print – is not emphasized strongly either, but is indicated indirectly in the distant landscape below. It is also worth pointing out that there is no carriage centrally placed in the picture, a mandatory feature, if the painting was primarily intended to show the improved quality of the road. While Vogt does make a point of showing the qualities of the road itself, the composition and effect of the painting actually seem to have been given a higher priority. The way he introduces the road into the picture and the way it leads us towards the view, testifies to this.

The fact that Anker commissioned a painting of Krokkleiva for his dining room at Bogstad demonstrates the significance of this particular road. The picture points towards his achievements as a road designer and reflects the importance Anker – an important figure of the Enlightenment in Norway – attached to an improved road system for the progress and future prosperity of the country. Although he also had personal interests in the area, it is not his economic activities that are shown in this painting, in contrast to the paintings of Vækerø and Bærum's Verk. The aesthetic aspects of the road and the landscape are emphasized to a greater extent in the Krokkleiva-painting. This can be seen in the way the road and the landscape are portrayed and used. It is to these aesthetical concerns that I will now turn.

VISITING KROKKLEIVA – VIEWING PRACTICES

So far I have only made a very brief mention of the figures in Vogt's *Krokkleiva*. Let us dwell on them for a while. What role do they play in the picture? Who are they and what do they do? How do they relate to and use the road and the landscape? And are there any connections between these figures and Anker's use of the road?

Unfortunately, the present state of the painting makes it difficult to see many of the details clearly. Still we can make out several figures (Figure 4.7). In addition to the earlier mentioned man with a horse and woman riding, it is possible to discern a dark-clad figure with a hat standing at the road with a dog behind him; apparently watching the small group who seems to be having a rest by the road-side. If we look further along the road we can also just about see some tiny figures – a man with a horse and carriage on their way downhill, and yet further away some tiny dots indicating other figures.

The man walking alongside his horse is the only figure showing a clearly utilitarian approach to the road. He is using it for transportation, and can be identified as a local peasant, dressed in a traditional white coat, breeches, knitted stockings with red bands, and a red cap. He is carrying a stick in his right hand and supporting the packsaddle with his left. The other figures are harder to identify and their purpose is less clear. Judging by their clothes, they seem to represent people with a more urban background. In the small party by the road a reclining man is leaning on his elbow, on his one side a woman is nursing a small baby, and straight behind him stands a man in a red jacket. His posture is active – as if he has just stood up or arrived – his left arm extended (pointing?) towards the road. A small dog is curled up in front of the group. Does the rather mysterious dark-clad man at the road

Figure 4.7 C.F. Vogt, Krokkleiva *(detail), ca 1812–1817, Bogstad Gård, Oslo (photo Arthur Sand)*

Reproduced with permission of Bogstad Gård.

belong to the same party? What about the woman on horse-back? How did the others get here? Should we imagine that horses and a carriage are waiting at Kleivstua – the travel lodge at the top of the road – and that the party has been walking down the road to look at the view? And if so, why are they all turned towards us, rather than shown watching the view? Does Vogt's painting allow such questioning or are the figures purely ornamental?

At least two of the other paintings in the dining room include identifiable figures: *Bærum's Verk* (Figure 4.5) and *Midsummer Eve at Bærum's Verk* (Figure 4.8).[23] Knowing this, it is tempting to speculate as to whether the figures in the painting of Krokkleiva might be identified too.[24] Could they, for instance, be members of the Anker family? Comparing them to identified figures in the two other paintings, none of them seem to match. However, even without being able to identify them, or establishing whether or not they were intended to be identifiable, their presence in the painting, and the way they relate to the road and the landscape, are significant. The figures are shown as being in the landscape and on the road with no particular purpose apart from just being there. Did the painter and his client, by including them, wish to make a statement about how the road and landscape could be used and seen? Could the inclusion of these particular figures even be inspired by actual visits to Krokkleiva? We know, in fact, that the Anker family visited Krokkleiva, and that on several occasions they brought distinguished guests there. Krokkleiva served as a tourist attraction and was subject to aesthetic appreciation right from the beginning. Two examples are particularly striking. In 1811 Count Reventlow, the Danish-Norwegian Prime Minister, visited Krokkleiva. The visit is described in his diary:

We travelled by boat across the lake Tyrifjorden to Krokkleiva, where a broad, but very steep road has been constructed. The view through a narrow pass bordered by tall, forest-clad rocks is very romantic and beautiful [...] In the evening we arrived at the ironworks at Bærum where the light from the blast furnace welcomed us.[25]

This visit was made only a few years before Vogt's picture was painted. Even without going into the details of his visit, it is clear that the road and the view made a strong impression on the Count. His description is short, but includes the main features that are repeated again and again in later accounts of Krokkleiva.

Several years later, in 1827, Lord and Lady Belgrave visited Bogstad.[26] Their stay in Norway also included a visit to Krokkleiva. Countess Karen Wedel-Jarlsberg (1789–1849), Peder Anker's daughter, arranged the tour and acted as their hostess.[27] Lady Belgrave describes Krokkleiva, which she mistakenly calls Kingrede, as 'the most tremendous we have seen; all but perpendicular, and very long'.[28] The two ladies left their carriage to be repaired at the top and 'walked down the hill, which forms a magnificent narrow defile, very precipitous and well wooded, through which, as you descend, a splendid view is gradually developed'.[29] After having dwelled upon the vista, Lady Belgrave goes on to describe the potential dangers of the descent: 'The hill was so steep that it was necessary

to tie together two wheels of Belgrave's carriage, and it then required the united strength of Zetterberg and himself to prevent its running over the horses'.[30] Keeping the figures in Vogt's painting in mind, it is worth noting that the Belgrave party did actually walk down Krokkleiva, and although this, judging by the above account, was to some extent a safety precaution, it is hard not to interpret it as a way of enjoying a better view as well.

My main point here is to show that Peder Anker and his family arranged visits to Krokkleiva in order to entertain guests, probably with the intention of showing them both the road and the view. Krokkleiva and its extraordinary view had been noticed earlier, most notably by the Danish painter Erik Pauelsen on his tour through Norway in 1788.[31] Still, it was only somewhat later in the nineteenth century that Krokkleiva gained its reputation as a tourist sight.[32] In the case of Vogt's painting and the Anker-family's use of the road, the practice of painting the road and the landscape, and the practice of viewing them come together in a particularly

Figure 4.8 The dining room at Bogstad Gård towards the west featuring Vogt's Midsummer Eve at Bærum's Verk (photo by Arthur Sand)

Note the view onto the garden, the lake Bogstadvannet and to Jens Juel's portrait of the Anker-family in the room next door. Peder Anker is wearing the uniform of his office as *Generalveiintendant* [Intendant General of Roads]. Reproduced with permission of Bogstad Gård.

striking way. This case represents an important moment in establishing Krokkleiva as a place for experiencing the landscape aesthetically. A practice that can be traced through countless later visits, written accounts and depictions, and that still continues today. It is probably no coincidence that the guests mentioned above came from abroad. They came from geographical and social backgrounds where the aesthetic appreciation of the landscape had a longer tradition. In this context one should also bear in mind Anker's educational background and especially his Grand Tour. Living and travelling abroad for several years in his youth, he became acquainted with European art, new ideas about the landscape and new technologies that must have been decisive for his later career and life.

BRINGING THE VIEW BACK HOME

Let us return to the dining room at Bogstad one last time. The inclusion of the painting of Krokkleiva points specifically to the pride Anker must have taken in the new road, but it also points towards a new way of using and experiencing the landscape. The painting might have been used to raise expectations before visiting the place itself, it might have been used to discuss a visit after it had taken place or it might have been used as a substitute for those not having the opportunity to visit Krokkleiva themselves. But it might also have been used for discussing issues concerning the introduction of new technologies and the state of the art of road building. Vogt's *Krokkleiva* can be seen as a prism for these different concerns.

By bringing the view back home, and letting his many guests see it, Anker contributed to making Krokkleiva better known, both as a place to visit and as a place to paint. This is confirmed by the fact that Vogt, probably just a few years after having finished the picture at Bogstad, painted a very similar version of the motif for King Carl Johan, and that the Queen and the King actually visited Krokkleiva some years later.[33]

The Krokkleiva-painting furthermore epitomizes the dialectics between representation and landscape that are at work when seeing Vogt's paintings together: Inside the dining room the painted views of Anker's estates merges with the real views onto his English garden at Bogstad, all overseen by the Anker family itself, grandly portrayed in the room next door.

ENDNOTES

1 After Denmark-Norway had been defeated in the Napoleonic wars, Denmark was forced to cede Norway to Sweden in 1814 (the Kiel Treaty). An overview on Norwegian history in English is given in Rolf Danielsen (et al.), *Norway: A History from the Vikings to Our Own Times* (Oslo: Scandinavian University Press, 1995).

2 Today Bogstad Gård is a museum (www.bogstad.no).

3 Krokkleiva was listed by the Directorate for Cultural Heritage in 1957 and remains a popular walking route.

4 Several earlier depictions of Norwegian landscapes are analysed in Brita Brenna's article in this book. Like the Danish artist Erik Pauelsen (mentioned by Brenna), Vogt received his education at the Art Academy of Copenhagen. Vogt came to Norway around 1810 and established himself in Christiania.

5 Count Herman Wedel-Jarlsberg (1779–1840) married Peder Anker's daughter Karen in 1807. He was the leading advocate of the Norwegian union with Sweden and became minister of finance in 1814.

6 With the exception of the chandelier and the chairs, which are later additions, the dining room has hardly been altered since the time of Peder Anker. Carsten Hopstock, *Bogstad. Et storgods gjennom tidende* (2 vols, Oslo: Boksenteret/Bogstad Stiftelse, 1997), vol. 2, p. 25.

7 The most relevant passages on the dining room paintings are found in: Carsten Hopstock and Holger Koefoed (et al.), *Billedkunst på Bogstad* (Oslo: Bogstad Stiftelse, 2003),

Yngvar Hauge, *Bogstad 1773–1955* (Oslo: Aschehoug, 1960), Bård Frydenlund, *Stormannen Peder Anker* (Oslo: Aschehoug, 2009) and in H.O. Christophersen, *Krokskogen i gamle dager* (Oslo: Grøndahl, 1981).

8   *Bærum's Verk*, 1813 and *Midsummer Eve at Bærum's Verk*, 1817. The Norwegian churchman Claus Pavels visited Bogstad in 1812 and mentions (in a diary entry dated 6 December) that he saw landscapes by Vogt, decorating one of the rooms at Bogstad. Claus Pavels and Ludvig Daae, *Claus Pavels' Dagbøger for Aarene 1812–1813* (Christiania: Foreningen, 1889), p. 75. These are the only sources for the dating of the paintings.

9   They both visited Bogstad during the 1790s, about 15–20 years before Vogt painted there. Wollstonecraft wrote rather critically about Anker's English garden and Malthus commented on his visit in his travel diaries. Mary Wollstonecraft, *A Short Residence in Sweden, Norway and Denmark* and William Godwin, *Memoirs of the Author of 'The Rights of Woman'* (London: Penguin, 1987), p. 146. Patricia James (ed.), *The travel diaries of Thomas Robert Malthus* (Cambridge: Cambridge University Press/The Royal Economic Society, 1966), pp. 104–5.

10   Hopstock, *Bogstad. Et storgods*, vol. 2, p. 48.

11   The dining room could accommodate up to 24 persons. When larger parties were present, they would dine in the hall, surrounded by Anker's art collection, most of which he probably bought at his Grand tour through Europe as a young man. Hopstock, vol. 2, p. 25.

12   The art historian Carsten Hopstock does to some extent pay attention to the roads in some of Vogt's paintings. However, he does not mention the painting of Krokkleiva.

13   The road was built around 1775, and is the predecessor of present-day Vækerøveien.

14   The English translation of the title is taken from Malthus, p. 105.

15   'skjønne, snorlike vei'. Cited from Just Broch, 'Norges Generalveiintendant Peder Anker', *Meddelelser fra veidirektøren*, nos 1 and 2, *Teknisk ukeblad* (1930), p. 1.

16   Hopstock, *Bogstad. Et storgods*, vol. 1, p. 239.

17   A part of the road had been built by the former owner of Bærum's Verk, Conrad Clausen (1754–1785), but Anker upgraded the road and extended it from Fossum (near Bogstad) to Maridalen.

18   The timber was floated down the lake Randsfjorden, via the river Randselva to Steinsfjorden and Åsa, then hauled up a high hill and landed in the waterways of Sørkedalen (north of Bogstad), eventually ending up at Anker's sawmills.

19   Anker was responsible for Akershus Amt. (The old name of the area that today include the counties of Oslo, Akershus, Østfold, Vestfold, Buskerud, Oppland and Hedmark.) For more information about the office and the appointment, see for example Frydenlund and Broch.

20   Routes down Krokkleiva have a long history and have been used since the Middle Ages as a part of the main route between Oslo/Christiania and Ringerike. Two alternative routes have existed: Nordkleiva [the northern route] in older times, and in more recent times, Sørkleiva [the southern route]. The road constructed by Anker is running down Sørkleiva. However, there was a road in Sørkleiva prior to this, traces of which still can be seen. It was made by Conrad Clausen and was used for transporting charcoal for his ironworks at Bærum's Verk. Christophersen, *Krokskogen i gamle dager*, pp. 52–60.

21   The French principle was introduced in Norway via Denmark, where the French road engineer Marmillod (1720–1786) was working. See for example Terje Bratberg, 'Veihistorisk skisse' in Ellen Margrethe Devold (et al.), *Vegvalg: Nasjonal verneplan: veger, bruer, vegrelaterte kulturminner* (Oslo: Statens vegvesen, Vegdirektoratet, 2002).

22   Hopstock, *Bogstad. Et storgods*, vol. 1, p. 108 and Frydenlund, *Stormannen Peder Anker*, p. 87.

23   In a small sketch, conserved at Bogstad, several of the figures in this last painting are named. Peder Anker

himself, his daughter, his son-in-law and about 20 other persons are identified in the sketch.

24 To my knowledge, no attempt has yet been made to identify these figures.

25 Chr. D.F. Reventlow, *Min reise i Norge 1811* (Oslo: Gyldendal, 1955), p. 40–41. Author's translation.

26 The couple might be better known as Richard Grosvenor, second Marquess of Westminster and his wife Elizabeth, Marchioness of Westminster. These were the titles that they received upon the death of Richard Grosvenor's father.

27 This visit took place several years after Vogt painted the picture and three years after Peder Anker died. Naturally it could not have been a source of inspiration for the painting, however it is relevant to note that the Anker family continued this practice.

28 Elizabeth Mary Grosvenor, *Diary of a Tour in Sweden, Norway, and Russia in 1827, With Letters* (London: Hurst & Blackett, 1879), p. 61.

29 Ibid., p. 61.

30 Ibid., pp. 61–62.

31 Pauelsen's paintings are the earliest known depictions of Krokkleiva. See Brita Brenna's article in this volume.

32 Margit Harsson, *Kongevegen over Krokskogen* (Røyse: Hole historielag, 1997), p. 29.

33 Magne Malmanger (ed.), *Malerisamlingen på Bygdø Kongsgård: Carl Johans samling av norsk kunst* (Oslo: Nasjonalgalleriet, 1982), pp. 33, 58. Sverre Grimstad, 'Kongens utsikt', *Ringerike* (Hønefoss: Ringerikes museum, Ringerike ungdomslag and Ringerike historielag, 2000), pp. 50–52.

5   Landscape's Imprint: On the Physiognomy of Plants around 1800

Charlotte Klonk

During the eighteenth century an astonishing number of new and unknown plants arrived in Europe on board of frigates sailing between the colonies and their mother countries. Lacking knowledge about their original habitat, botanists explored the physiognomy of plants as a way to deduce the nature of their native growing conditions in the belief that their outward appearance might betray clues about their inner needs. The study of physiognomy had been established in Antiquity, but its subject had traditionally been the human face. Its radical growth in popularity around 1800, however, made it possible to also think about plants and their settings in terms of physiognomy. Three different types of plant physiognomy emerged, I will argue here, reflecting the conflict between art and science around 1800: firstly, there was a vitalist explanation, secondly, a metaphysical conceptualization and, thirdly, an empirical approach. All three stemmed from a field of inquiry that seems far removed from us today; in it, scientific findings and aesthetic experience, political convictions and personal circumstances were closely intertwined. These three different ideas about plant physiognomies also expressed fundamental changes in landscape art occurring after 1800. Travel played a key role therein and has been little acknowledged thus far.

I will begin with a general introduction to the study of physiognomy and landscape backgrounds around 1800. Then, I will turn to the botanist Robert John Thornton. Thornton had never been abroad, therefore, he conceived of foreign plants in a familiar natural habitat that also conformed to established European aesthetic convictions. Others, like the German Romantics, I will move on to argue, began to study the natural setting of native vegetation more closely in the hope of transcending the specific for the universal. At the same time, artists who accompanied explorers – James Cook, for example – introduced yet another format for nature's depiction.[1] The contribution made to this by the German naturalist Alexander von Humboldt, the focus of the third part of this article, has so far been largely neglected. In 1804, shortly after returning from five years in Latin America, he began to write about the physiognomy of foreign plants in the tropics. Humboldt's text inspired a number of landscape artists, bringing about, as I will conclude, a new conception of nature.[2]

FACES AND LANDSCAPES

More than any other figure, it was the Swiss physician and vicar Johann Kaspar Lavater who was responsible for the sudden popularity of the study of character at the end of the eighteenth century. In his work, the *Fragments of Physiognomy for the Promotion of Human Understanding and Human Love*, published between 1775 and 1778, Lavater attempted to reconcile ancient ways of thinking about physiognomy with the new systematic science of his day. Lavater used silhouettes to measure the structure of the face and to collect data suited to comparison and differentiation. Subtle differences, such as the tapering of the nose, the height of the forehead, the line of the eyebrows and the shape of the mouth, were all given great significance. According to Lavater, such details revealed the character of the individual more than words ever could. In the second volume of his *Fragments of Physiognomy*, Lavater wrote that 'one can detect nine horizontal main sections in each silhouette':

1. The incline from the parting of the hair to the hairline. 2. The contours of the forehead to the eyebrows. 3. The space between the eyebrows and the base of the nose. 4. The space between the nose and the upper lip. 5. The upper lip itself. 6. The lips. 7. The upper chin. 8. The lower chin. 9. The neck, including the back of the head and the nape of the neck.

He continued:

Each of the parts of these sections is like a letter, sometimes a syllable, often a word or even a sentence. It tells the truth. When all sections are in harmony, they reveal the character of the person in such a way that [even] a peasant or a child can discern it in a silhouette.[3]

Figure 5.1 *Joseph Karl Stieler,* Ludwig van Beethoven composing the Missa Solemnis, *oil on canvas, 1820*
Reproduced with permission of the Beethoven-Haus, Bonn.

For Lavater, the study of the face was no longer about recognizing individual features as the manifestations of particular kinds of characters, as had been common in physiognomy since Antiquity. Instead, it was about understanding the whole personality. In order to do this, many elements had to be brought into play. Together they were to create a clear and concrete image; Lavater called it a total impression.

A portrait painted between 1821 and 1822, now in the collection of the Beethoven-Haus in Bonn, demonstrates what this 'total impression' entailed. It depicts Ludwig van Beethoven, whose fame as a composer was already well established (Figure 5.1). Even while alive, Beethoven was considered an artistic personality who defied conventions. His emotive art was believed to reflect his own passionate experiences and, in so doing, to express universal human concerns too. To fully appreciate this portrait, one must also know, however, that Beethoven was seen as the first composer to try to apply his artistic concepts without having been commissioned to do so. As a result he became the epitome of the independent artistic genius. In both posture and facial expression, Joseph Karl Stieler's rendering of Beethoven signals the composer's intention to confront his inescapable fate with unfaltering resolve. He appears as a stern-faced, middle-aged man, dressed in a morning gown and with conspicuously unkempt hair: in short, a person with every appearance of not heeding the dominant conventions of his day. In the picture Beethoven is immersed in his own thoughts; we see him in the act of composing the *Missa Solemnis* jotting down his genial creative impulses. Lavater would have perceived genius in him immediately. With their bushy eyebrows and upward gaze, his deeply-set eyes reveal, according to Lavater's theory, 'a stroke of genius'. Moreover, Beethoven's exposed forehead, gently curving to the hairline, suggests a 'nevertheless altogether rational' character.[4] His pursed lips demonstrate precisely that uncompromising power of will one associates with an independent mind. But, it is only through, to use Lavater's phrase, the picture's total impression – that is the complete display of Beethoven's individual characteristics – that his uniqueness and unmistakable personality is made apparent.

Yet, there is one element not featured in Lavater: the relationship between the background and the portrayed face. In the painting, we are also presented with a wild, uncultivated view of nature. The background adds a further layer of meaning to the portrait. It seems to be saying that just as nature has no bounds, Beethoven's lack of artistic inhibition and restraint manifests itself in the *Missa Solemnis*.

Analogies of this kind were, of course, nothing new in the art of portraiture. Nearly two centuries before, Anton Van Dyck had depicted the English King, Charles I, on horseback in front of an ancient gnarled oak tree in order to symbolize his deeply-rooted, ancient right to the English throne. Around 1800, a strong belief that the living being was dependent on and shaped by its surroundings emerged from this tradition; it was a more inclusive view and we would perhaps describe it as ecological today. It had its origins in botany or, to be more precise, in the physiognomy of plants.

A VITALIST CONNECTION

The first botanical treatise to show plants integrated within their natural surroundings was Robert John Thornton's *A*

New Illustration of the Sexual System of Carolus von Linnaeus published between 1798 and 1807. It illustrated Linnaeus, however, in a superficial sense only. Financed through subscriptions and produced over the course of nearly ten years, Thornton published a series of large-scale, hand-coloured etchings and aquatints in a section entitled *Temple of Flora: Or, Garden of Nature*. Showing plants in landscape settings, these plates radically differed from previous botanic illustrations, such as those in Georg Dionysius Ehret's

Figure 5.2 Philip Reinagle, The Superb Lily, *mezzotint printed in colour*

Robert John Thornton, *The Temple of Flora* (London, 1807). Private collection.

Plantae Selectae, which appeared between 1750 and 1773. Thornton had engaged the artists who executed the images and supervised them closely. One, Philip Reinagle, the artist who designed among others the plate of the lily, was already well known to his contemporaries as a painter of animals and landscapes.

He followed Thornton's instructions and depicted the plant before a mountain landscape which stands in stark contrast to the flower's graceful form (Figure 5.2). Yet neither Reinagle nor Thornton had encountered the plant in its native setting. The *Superb Lily* is a native of eastern North America and had only recently been introduced to the temperate climate of British gardens. Thornton obviously imagined its natural setting to be much less benign. Magnified and truncated, there is something heroic about the colourful lily now set against the dark, awe-inspiring background. Just as stormy seascapes and barren landscapes often characterize the subjects of Joshua Reynolds' portraits as legendary seafaring heroes, so too the landscape backgrounds in Thornton's publication serve as a characterization of the plants featured in them.[5]

Thornton's depictions range from the sublimely heroic, as in the *Superb Lily* or the *Dragon Arum,* to the sweetly beautiful. In the rose plate (Figure 5.3), he ordered the depiction of a temple at sunrise in a reference to the then-popular, classical landscape paintings of the French seventeenth-century artist Claude Lorrain. He intended viewers to be reminded of the ideas of eternal beauty associated with Lorrain's paintings and to transfer these, by analogy, to the flower and its perceived character. The landscapes in Thornton's plates never present the plants in their true natural habitat. Indeed, sometimes the flowers even appear in a setting in which they would never naturally grow as, for example, with the highly-artificial combination of snowdrops and crocuses high up on snow-covered Nordic hills.[6] In fact, Thornton's approach in the *Temple of Flora* can be seen as thoroughly anthropomorphic: the illustrations are portraits whose subjects are plants. As in Stieler's portrait of Beethoven, there seems to be an unbroken ontological continuum between setting and subject, appearance and perception. On the one hand, Thornton's landscapes determine the flower. On the other, in the absence of any knowledge about the flower's true habitat, the plant's external appearance was believed to hold the key to the nature of its natural setting. It is, therefore, not the original habitat which is depicted but rather the shared aesthetic character.

Thornton's plant portraits showed a new understanding of environmental effects on the character of life forms;

Figure 5.3 Robert John Thornton, Roses, *mezzotint printed in colour*
Robert John Thornton, *The Temple of Flora* (London, 1807). Private collection.

they also revealed, perhaps more importantly, a yearning for classification not along Linnean lines but according to aesthetic perceptions. It took the next generation of English and German landscape painters before the interdependence of nature's geographical, geological, meteorological and vegetable manifestations began to be more closely studied. They did this without leaving their respective countries and, significantly, foreign plants held no appeal.

METAPHYSICAL LANDSCAPES

Like John Constable in Britain, Caspar David Friedrich never left his native Germany. In his paintings of German oaks, the trees appear exactly as one sees them in the Nordic landscape. They are solitary forms beside waterholes where wild boar wallow and feed on the fallen acorns. Generations of scholars have argued that Friedrich sought to create a meaning beyond that which is immediately visible. In

Figure 5.4 Carl Gustav Carus, Frühlingslandschaft im Rosenthal bei Leipzig, *oil on canvas, 1814*
Reproduced with permission of Galerie Neue Meister, Staatliche Kunst-sammlungen Dresden.

Friedrich's gnarled oak trees, some have even gone so far as to see references to Germany's political situation following the French occupation and the subsequent Restoration, which had disappointed many by capping the burgeoning patriotic liberation movement.[7]

One can see similar symbolic meanings in the pictures of Friedrich's one-time friend, Carl Gustav Carus, who, in his faithful rendering of nature, achieved a level of meaning that transcended the actual depicted view. Carus was a doctor and natural philosopher and in his eighth 'landscape letter' he explicitly called for a physiognomic approach to painting.[8] Indeed, even in the earliest of his works, the skill with which he transferred exact observations to the canvas is remarkable. In a painting entitled *Spring Landscape in Rosenthal near Leipzig*, from 1814, a stream winds through a bright wetland forest (Figure 5.4). The trees are only just starting to bear leaves and the low-growing plants on the wet grassland are just beginning to blossom. As Stefan Grosche has noted, the nature of the wood changes between the fore- and backgrounds:

While, in the picture's foreground the hard and seemingly lifeless contours of the branches cut into the reflected surface of the still water abruptly, our gaze is drawn to the spring atmosphere of the background by the diagonal row of trees. In the foreground, bare alder trees determine the line of sight. However, in the background, the softer curves of the already-budding willow branches dominate the picture.[9]

The *Spring Landscape in Rosenthal near Leipzig* would remain important to Carus in later life, when, in his letters on landscape painting, he expressed the effect he had striven for as follows:

The feeling of exertion, of encouragement, of progression, the feeling of true inner clarity and peace, [...] are in this

case also the four stages, to which spiritual life with all its unceasing variety can be traced.[10]

For Carus, as for Friedrich, the depiction of plants in their natural habitat was a way to understand life's more fundamental processes.

AN EMPIRICAL OUTLOOK

Where Thornton saw an overriding vitalist principle at work,[11] Friedrich and Carus sought *meta*physical forces lying beyond the visible world. For them, nature, in its various forms, drew one into the realm of ideas. Alexander von Humboldt, in contrast, was not attracted by such leaps of faith. Science, for Humboldt, took place against a backdrop of transparent processes, effects and influences and, in his correspondence with Carus the distance between the two men is obvious.[12] Humboldt had been interested in a different type of scientific study long before he boarded the *Pizarro* in La Coruña in Spain on 5 June 1799. His very first publication on fungi and lichens was an exercise in Linnaean taxonomy.[13] In it, he had also acknowledged the need for metaphysical speculation although he later wrote self-critically to Schiller:

> In the way that natural history was practiced until now, where one has simply clung to the differentiation of forms, the study of the physiognomy of plants and animals, the learning of characteristics and how to recognize them, and confused all this with sacred science itself, botany was never to have been an object of contemplation for men of ideas.[14]

Here, the term physiognomy appears for the first time in Humboldt's writings but with entirely negative connotations. Physiognomy describes the mere registration and classification of types in the Linnean tradition as Humboldt had practiced it until then.

Humboldt advanced two alternative strategies: firstly, he suggested, understanding the sensory effects of the plant kingdom on the human mind; secondly, he proposed, a study of the spread and distribution of plants over the earth's surface. Neither was, however, evident in Humboldt's practice in the field. His travel diaries from the period are filled with observations, reports, measurements, scientific notes and minor discourses. The study of aesthetic effects, so important to physiognomy, and which Humboldt had called for in his letter to Schiller was only a subordinate concern. Things changed when he returned from the tropics five years later. Confronted by the Berlin winter, the Napoleonic occupation and by the decline of Prussian scientific activity that had taken place in his absence, Humboldt yearningly cast his mind back to a richer, more fertile world. It is worth quoting him at length to show what role aesthetic considerations now played in his thinking:

> In the tropics vegetation is generally of a fresher verdure, more luxuriant and succulent, and adorned with larger and more shining leaves, than in our northern climates. The "social" plants, which often impart so uniform and monotonous a character to European countries, are almost entirely absent in the tropical regions. [...] The great elevation attained in several tropical countries, not only by single mountains but even by extensive districts, enables the inhabitants of the torrid zone – surrounded by palms, bananas, and the other beautiful forms proper to those latitudes – to behold also those vegetable forms which, demanding a cooler temperature, would seem to belong to other zones. [...] Thus it is given to man in those regions to behold without quitting his native land all the forms of vegetation dispersed over the globe, and all the shining worlds which stud the heavenly vault from pole to pole. These and many other of the enjoyments which Nature affords are wanting to the nations of the North. [...] Individual plants languishing in our hot-houses can give but a very faint idea of the majestic vegetation of the

tropical zone. But the imitative art of the painter opens to us sources whence flow abundant compensations, and from whence our imagination can derive the living image of that more vigorous nature which other climes display. In the frigid North, in the midst of the barren heath, the solitary student can appropriate with the help of pictures mentally all that has been discovered in the most distant regions, and can create within himself a world free and imperishable as the spirit by which it is conceived.[15]

Somewhat problematically, however, the pictures supposed to be the source for Humboldt's mental escapism did not yet

Figure 5.5 Franz Post, Brazilian Landscape with Ant Bear, *oil on wood, 1649*
Reproduced with permission of the bpk/Bayerische Staatsgemäldesammlungen, Munich.

exist. So far, no artist had managed to convey a sense of the richness of the tropics in pictures, as Humboldt himself had done in words in his 1807 European bestseller, the *Aspects of Nature*. Humboldt singled out Franz Post, a painter from Harlem, as being the first to have tried; Post had accompanied Moritz von Nassau, the commander of the Dutch West Indies Company, to Brazil between 1637 and 1644. Nonetheless, Post's pictorial compositions are still quite conventional (Figure 5.5). In the foreground, individual exotic plants are lined up as if in a herbarium. They are so clearly defined that they could be cut outs. Meanwhile, the wide expanse of the background is left largely atmospheric and unspecific. Here, there is no trace of what Humboldt was really after: a physiognomy of the tropics.

By 1807, the term 'physiognomy' had lost all negative connotations for Humboldt. In a text on plant physiognomy that he published in *Aspects of Nature*, Humboldt wrote that everything was reducible to the 'total impression', to that which determines the character of the vegetation and thus the viewer's spirit.[16] The few depictions of landscape that did feature in Humboldt's own work were either his own renderings and were, as such, quite diagrammatic, for example the famous 1807 plate showing the distribution of plants according to elevation in the Andes, or they had been created by artists, like Joseph Anton Koch, who had not been to the tropics. One etching based on a work by Koch (himself highly revered by Humboldt) was entitled *Vues des Cordillères* and was printed in Humboldt's travel account from 1810–1813. It illustrates the problem clearly (Figure 5.6). Seen from an elevated perspective, a pass in the Andes leads down into a valley surrounded by a chain of volcanoes. The shrubs and trees in the etching are rendered relatively schematically and could just as easily adorn a European alpine view. The blooming agave to the right of the picture is a sole indication that the scene is foreign. Humboldt's scientific preoccupations, and the ideas he outlined to the Academy of Sciences in Berlin on 30 January 1806 in his lecture 'Ideas towards a Physiognomy of Plants', are not

Figure 5.6 Joseph Anton Koch, Passage du Quindiu, *etching*
Alexander von Humboldt, *Vues des Cordillères, et Monumens des Peuples indigènes de l'Amérique* (Paris, 1810[-1813]). Private collection.

apparent here at all. The picture does not convey a total impression of a landscape, determined, first and foremost, by the dominant plant forms. Moreover, Humboldt devised, in a text also conceived in 1807, seventeen main forms or types of flora, 'whose study [would] be of particular importance to the landscape painter'.[17] The list began with banana and palm trees and went all the way down to the insignificant lichens and fungi that he had observed in the mineshafts around Freiberg. Humboldt was aware that by devising such a list he was developing a typification that differed from that of Linnaeus. His predecessor had given importance to the almost invisible aspects of plants that helped them to reproduce. In contrast, Humboldt focussed on the view of the landscape as a whole.

In the years that followed, Humboldt's fame and the popularity of his writings inspired flocks of researchers, artists and travellers to venture to the tropics. Their efforts to record the physiognomy of plant groups according to Humboldt's scheme are obvious. The resulting pictures had nothing in common with Thornton's plant physiognomies or the paintings of the German Romantics.

COCOS

botryophora .   schizophylla .

Figure 5.7 Carl Friedrich Philipp von Martius, Cocos botryphora.
Cocos schizophylla, *coloured lithograph*
Carl Friedrich Philipp von Martius, *Historia naturalis palmarum*, vol. 2
(Munich, 1823–1832). Private collection.

Carl Friedrich Philipp von Martius was one of the first
explorers to follow in Humboldt's footsteps. In 1817, he
was instructed by the Bavarian king to embark, together
with the zoologist Johann Baptist Spix, on an Austrian-
led expedition to Brazil.[18] Four years later, and after
experiencing unspeakable hardships, they finally returned
to Munich. In 1823, Martius set about publishing an
exceptionally detailed and elaborately illustrated treatise

on palms. It is no coincidence that Martius chose to
concentrate on this particular plant form; in Humboldt's
text on the physiognomy of plants in the 1808 *Aspects of
Nature*, he had asserted that palms were 'the noblest of all
vegetable forms'.[19] Included alongside detailed illustrations
of isolated plants and their significant components, an
array of impressively coloured lithographs of various
species in their natural habitat appeared in volumes 2 and
3 of Martius' treatise. This treatise was probably the first
botanical work, with the exception of Thornton's *Temple
of Flora*, to also include landscape pictures.[20] Martius was,
however, no landscape artist. For several of the illustrations,
he used landscapes by Franz Post known to him from a
Munich collection. For others, he turned to Johann Moritz
Rugendas, a contemporary painter from Augsburg, favoured
and supported by Humboldt.

In Plate 95 from the second volume of the *Historia
naturalis palmarum*, one can see the palm species *cocos
botryphora* and *schizophylla* in the foreground; a section
from Post's 1649 *Brazilian Landscape with Armadillo* (then
part of the collection of the Schleißheimer Galerie in Munich
and today in the Old Pinakothek) has been used for the
background (Figures 5.5 and 5.7). Yet where, in Post's pictures,
the contrast between the overly exact positioning of plants
in the foreground and the blurred forms of the background
failed to create a coherent impression of the landscape, the
opposite is true of Martius' work. The most striking of plant
forms, the palms, are depicted naturally and in proximity
to smaller species such as cacti and bananas; meanwhile, in
the background, the rendering of the uniform surface of the
canopy of leaves has been kept deliberately vague. Humboldt
had stated in *Aspects of Nature* that plants could not be
closely delineated from a distance and this is how we see them
here. Thus, Martius created an exemplary body of illustrative
material for Humboldt's empirically orientated physiognomy
of plants. His *Historia naturalis palmarum* rendered the
global history of nature and geography as impressively visible
as Humboldt had envisaged it.

Figure 5.8 Johann Moritz Rugendas, Campos, *lithograph*
Johann Moritz Rugendas, *Malerische Reise in Brasilien*, vol. 1 (Mülhausen, 1827). Private collection.

A similar approach was taken by Rugendas, who Humboldt would go on to call the 'pioneer and father of all art in the depiction of the physiognomy of nature'.[21] During his two journeys to Latin America, Rugendas was considerably occupied with attempts to capture the phenomena of light and mass and their distribution on the canvas. In the magnificent oil sketches he made on the spot, landscape physiognomy, in Humboldt's sense of the word, was not a primary concern. In his published works, however, Rugendas was more aligned with Humboldt who was, after all, his sponsor and advocate. Each and every one of the lithographed landscapes in the work dedicated to his Brazilian travels, and dating from 1827 to 1835, adheres to a clear physiognomy dominated by a specific plant type. This holds true for both the desert-like mountain landscapes, where nothing but agaves grow, and the expansive landscape panoramas dominated by palms (Figure 5.8).

Rugendas also contributed rich jungle views to Martius' work. These give a sense of the wealth of plants and flowers that had moved Humboldt to declare that the physiognomy of

LEPIDOCARYUM gracile. SAGUS taedigera.

Figure 5.9 Johann Moritz Rugendas, Lepidocaryum gracile, Sagus taedigera, *coloured lithograph*
Carl Friedrich Philipp von Martius, *Historia Naturalis Palmarum*, vol. 2 (Munich, 1823–1832). Private collection.

vegetation below the equator 'boast[ed] of greater size, majesty and diversity than in the temperate zone' (Figure 5.9).[22] For Humboldt, tropical abundance encapsulated a part, however small, of the original whole, of the earth's primeval vegetation, which over the course of several millennia had been slowly destroyed in other regions by human intervention. Unlike Carus, Humboldt did not see an aesthetic approach to nature as an end in itself but rather as an aid to scientific advancement.

It was not the pictures *per se* that made it possible to 'recognize the inner, secret workings of the forces of nature' but rather the study of the distribution of vegetation over the earth's surface. Humboldt believed that science allowed one 'to draw conclusions about the future and to predict the recurrence of major natural events'.[23]

It becomes clear, here, that to call Humboldt a Romantic would be mistaken. It is true that he strove to discover the all-illuminating primeval form underlying all vegetation. But he did not look in places other than in the outward forms of plants. He did not believe in an invisible vitalist principle, as Thornton did, or in metaphysical ideas, as Carus and Friedrich did. Instead, Humboldt immersed himself in empiricism. Contrary to what scholars have repeatedly asserted about Humboldt's relationship with the Romantics, he appears to have been fully aware that the relationship between natural science and the sensory pleasure derived from an aesthetic view of nature, such as physiognomy conveyed, would always be fraught. In his *Ideen zu einer Geographie der Pflanzen*, written as he conceived, and parts of which perhaps even before, his Berlin lecture on the physiognomy of plants the fusion of observation and measurement, experience and experiment, enjoyment of nature and its mastery is palpable.[24] The first sixth of the text was dedicated to the importance of aesthetics; the remaining five-sixths were filled with measurements, observations, experiments and tables all of which served scientific and practical questions or aims. Humboldt's landscapes were not painted with a painter's palette but with the instruments and measuring apparatus of a scientist. In the process, Humboldt united physiognomy and empiricism and laid the foundations of modern ecological inquiry.

THE POLITICS OF PHYSIOGNOMY

When I began, I emphasized the popularity of the study of physiognomy around 1800. This is worthy of note,

particularly since, in all fields, there are concepts that have been laid to rest once their significance has become redundant. Physiognomy has not been one of these. Known since Antiquity, interest in it grew in the eighteenth century and physiognomy experienced a revival, not only in the 20s and 30s of the last century, but also in the ecology movement of the 70s and 80s.[25] One reason for its longevity might be physiognomy's dependence on aesthetics. As aesthetic principles have changed, so too the aspects taken into account in the study of physiognomy have been reconfigured. Around 1800, three different types of aesthetic conviction allowed for the emergence of three corresponding kinds of plant physiognomy. Thornton proposed a vitalist connection between plants and their environment; Friedrich and Carus saw metaphysical implications; and Humboldt attempted an empiricist account.

This also means, however, that, as a theory of values, physiognomy is heavily influenced by prevailing beliefs; that is to say, by the fears and ideas of those who draw on it. According to Robert John Thornton, the snowdrop did not lament its lowly status but the lily rebelled with disastrous consequences; when he emphasized that each plant has its own place in the natural order he surely also wrote with the contemporary political situation in mind.[26] Drawing an analogy between the natural order of the plant kingdom with that of the human realm, is a clear expression of his support for the current political order that had come under pressure in England and elsewhere in the wake of the French Revolution. Political motivations also underlay Humboldt's enthusiasm for the tropics. When Humboldt pitted the vegetal diversity of the southern hemisphere against the tiring uniformity of the northern, he was subtly celebrating anarchic individualism. It is, perhaps, no coincidence that he emphasized in his writing that 'no plant exerts dominance over the others in the rainforest' as Napoleon marched his way across Europe.[27]

Even the misuse of physiognomy in the racial ideology of the Nazis has not been able to bring about an end to its study. Like Thornton, but certainly far less benignly, the Nazis used physiognomy to detect foreignness among the native population. In fact, physiognomy is currently experiencing a revival both in esoteric new-age circles and in the empirically-based research on emotions by Paul Ekman, an American psychologist and anthropologist.[28] Humboldt sought a sympathetic world far away; Ekman, too, has found reassuring familiarity among the Fore tribes people in Papua New Guinea. As different as the theories are, physiognomy is a response to a mobile world in each case. As long as strange people and foreign things continue to encounter one another in an increasingly dense network of routes and roads, physiognomy will continue to serve a vital purpose.

ENDNOTES

1   See, for example, Geoff Quilley and John Bonehill (eds), *William Hodges, 1744–1797: The Art of Exploration* (New Haven and London: Yale University Press, 2004), and Harriet Guest, *Empire, Barbarism and Civilisation: Captain Cook, William Hodges and the Return to the Pacific* (Cambridge: Cambridge University Press, 2007).

2   A first attempt to chart Humboldt's influence has recently been presented in an exhibition in Berlin, see Sigrid Achenbach, *Kunst um Humboldt: Reisestudien aus Mittel- und Südamerika von Rugendas, Bellermann und Hildebrandt im Berliner Kupferstichkabinett* (Munich: Hirmer, 2009). Also: Renate Löschner, 'Humboldts Naturbild und seine Vorstellung von künstlerisch-physiognomischen Landschaftsbildern', in Karl-Heinz Kohl (ed.), *Mythen der Neuen Welt: Zur Entdeckungsgeschichte Lateinamerikas* (Berlin: Frölich & Kaufmann, 1982).

3   Johann Caspar Lavater, *Physiognomische Fragmente zur Beförderung der Menschenkenntnis und Menschenliebe* (vol. 2 [1776], Zurich: Orell Füssli, 1968), pp. 96–97. Unless otherwise stated, all translations are mine.

4   Lavater, *Physiognomische Fragmente*, vol. 4, pp. 84–89.

5   For a discussion of this comparison, see Charlotte Klonk, *Science and the Perception of Nature: British landscape art in the late eighteenth and early nineteenth centuries* (New Haven: Yale University Press, 1996), p. 60.

6   Robert John Thornton, *The Temple of Flora: Or, Garden of Nature. Part 3 of 'A New Illustration'* (London: Bessley, 1807), Plate III.

7   Berthold Hinz, Hans-Joachim Kunst, Peter Märker et al. (eds), *Bürgerliche Revolution und Romantik: Natur und Gesellschaft bei Caspar David Friedrich* (Gießen: Anabas Verlag, 1976).

8   Carl Gustav Carus, *Neun Briefe über Landschaftsmalerei, geschrieben in den Jahren 1815 bis 1824*, ed. Kurt Gerstenberg (Dresden: Jess, 1927), p. 61.

9   Stefan Grosche, 'Lebenskunst, Krankheitskunst, Heilkunst. Novalis in der Medizin von Carl Gustav Carus', in Petra Kuhlmann-Hodick, Gerd Spitzer and Bernhard Maaz (eds), *Carl Gustav Carus. Wahrnehmung und Konstruktion. Essays* (München: Deutscher Kunstverlag, 2009), p. 45.

10 Quoted after Kuhlmann-Hodick, Spitzer and Maaz (eds), *Carl Gustav Carus*, p. 36.

11 This is more fully argued in Klonk, *Science and the Perception of Nature*, pp. 60–65.

12 See Ingo Schwarz, 'Carus und Alexander von Humboldt', in Kuhlmann-Hodick, Spitzer, Maaz, *Carl Gustav Carus*, pp. 333–337. Only parts of his diaries are published: Alexander von Humboldt, *Reise auf dem Rio Magdalena, durch die Anden und Mexico*, vol. 1: *Texte* (ed.), Margot Faak (Berlin: Akademie-Verlag, 1986).

13 Alexander von Humboldt, *Florae Fribergensis specimen* (Berlin: Rottmann, 1793).

14 Letter to Schiller (August 6, 1794) in: *Die Jugendbriefe Alexander von Humboldts. 1787–1799*, ed. Ilse Jahn and Fritz G. Lange (Berlin: Akademie-Verlag, 1973), p. 364.

15 Alexander von Humboldt, *Ansichten der Natur*, ed. Hanno Beck, study edition 5 (Darmstadt: Wissenschaftliche Buchgesellschaft, 1987), pp. 191–192. Quoted from the

English translation that appeared in 1849 (Alexander von Humboldt, *Aspects of Nature*, trans. Mrs. Sabine, 2 vols. [London: Longman, 1849], pp. 245–46).

16 Alexander von Humboldt, *Aspects of Nature*, p. 234

17 Alexander von Humboldt, *Schriften zur Geographie der Pflanzen*, Hanno Beck (ed.), vol. 1 (Darmstadt: Wissenschaftlige Buchgesellschaft, 1989), p. 64.

18 Ludwig Tiefenbacher, 'Die Bayerische Brasilienexpedition von J.B. Spix und C.F. Ph. Martius 1817–1820', in Jörg Helbig (ed.), *Brasilianische Reise 1817–1820. Carl Friedrich von Martius zum 200. Geburtstag* (Munich: Hirmer, 1994), pp. 28–51.

19 Humboldt, *Aspects of Nature*, p. 238.

20 See Hans Walter Lack, *Ein Garten Eden. Meisterwerke der botanischen Illustration* (Köln: Taschen, 2001), p. 398.

21 Löschner, 'Humboldts Naturbild und seine Vorstellung von künstlerisch-physiognomischen Landschaftsbildern', p. 251.

22 Humboldt, *Schriften zur Geographie der Pflanzen*, p. 65.

23 Ibid., p. 66.

24 Alexander von Humboldt, *Ideen zu einer Geographie der Pflanzen* (Tübingen and Paris: Cotta/Schoell, 1807).

25 See Claudia Schmölders, *Das Vorurteil im Leibe: Eine Einführung in die Physiognomik* (Berlin: Akademie-Verlag, 1995), and Claudia Schmölders (ed.), *Gesichter der Weimarer Republik: eine physiognomische Kulturgeschichte* (Cologne: DuMont, 2000).

26 Robert John Thornton, *The Politician's Creed, Being the Great Outline of Political Science* (London: Cox, 1795), p. 26.

27 Humboldt, *Ansichten der Natur*, p. 50.

28 See, for example, Paul Ekman, *Emotions Revealed* (New York: Times Books, 2003).

6   Travel, *en route* Writing, and the Problem of Correspondence

Charles W.J. Withers

Travel writing and narratives of exploration – writing about routes and about route making – have been widely scrutinized.[1] Historians of science stress the importance of voyages of exploration and their printed narratives to the emergence of modern science.[2] Geographers have studied the practices of printed inscription central to institutions facilitating travel, exploration and trade and scrutinized the epistemic practices of geographical writing.[3] Book historians and others have shown the value of spatial and visual perspectives in understanding the making, distribution, and reading of printed texts.[4] Literary scholars have reviewed the connections between narratives of travel, empire, 'self' and 'other' and literary form.[5] Taken together, these scholarly concerns show that engagement with narratives of exploration and travel writing is an important interdisciplinary endeavour.

Important but incomplete: much work either focuses upon the content of travel narratives, to the neglect of epistolary conventions, or assumes as largely unproblematic the relationship between writers' narratives, their experiences in the field, and the printed version of their work. Because this is so, more remains to be known of the epistemic bases to authors' claims about the truth of their experiences, about route writing and travel narrative as a practice, and how authors and publishers established correspondence between what they saw or were told, and what they wrote about in the published accounts. Making the transition from the voyage to a narrative about it – the 'voyage into print'[6] – was not always easy. Publishers could and did alter their authors' manuscript narratives in order to serve different demands.[7] Authors or their interlocutors regularly modified their work prior to its publication, to say nothing of alterations in later editions. And there is a further issue here: namely, route writing and the route to truth. Because narratives about routes and route making did not always correspond to the printed versions, interrogating the embodied practices which facilitated travel narratives and investigating the relationships between narrative and practice become central to understanding the ways in which authors sought to establish a correspondence between their experiences of exploration and the textual representation of them.

Correspondence is a matter of epistolary culture and an epistemic desideratum: simply, that what is written about should correspond in some way to the world thus described. Epistolary practice and other forms of authorial convention and restriction are vital to narrative form. In epistemological terms, correspondence is often presumed: as a question of one's own eye witnessing, the testimony of reliable informants, or the truth you reason for yourself. Knowledge in the form of explorers' narrative is commonly based upon testimony and testimony upon trust – in one's self, one's instruments or in others' credibility.[8] Since publishers could alter their authors' words, and authors had to rely upon others' words, correspondence between the account as written and the account as published cannot be assumed and so must be tested. There is a third sense, too, in which we may understand correspondence: edition history. In numerous travel narratives, authors/publishers amended their works to produce new editions as new knowledge was brought to bear. In these issues of epistolarity, epistemology and edition history lie important questions about route making, route writing, and to the route to truth in travel accounts.

These connections have been addressed by one author in what he terms *en route* writing, a four-stage linear model from field notes, to the fuller journal, to the draft manuscript, and the published version.[9] In looking at questions of inscription, authorial practice and epistemology in narratives of exploration, this chapter addresses the 'possibility that closer attention to the practices, techniques and technologies of writing itself can show how they give shape to the meanings of geographical knowledge'.[10] What follows is rooted in the contention that more remains to be known of the epistemic bases to authors' claims about the truth of their experiences in their narratives.

My particular focus in illuminating these issues involves the narrative work of Captain George Francis Lyon (1795–1832). Lyon was born in 1795, entered Britain's Royal Navy in 1808 and was commissioned lieutenant in 1814. He took part (as a volunteer) in Joseph Ritchie's trans-Sahara expedition between early 1819 and March 1820. Promoted commander in 1821, Lyon (in command of *HMS Hecla*) sailed with Captain William Edward Parry (captain of *HMS Fury*) as part of Parry's second voyage, between 1821 and 1823, in search of that peculiarly iconic route, the north-west passage. In January 1824, Lyon was given charge of *HMS Griper* (which had been on Parry's first voyage of 1819) with the brief to reach Repulse Bay in the American North and continue the survey of the mainland coast where Sir John Franklin had ceased. Lyon was unsuccessful in this. In 1825, Lyon was in Mexico as an inspector for English mining concerns there. Upon return to Britain in January 1827, most of Lyon's papers and specimens were lost as his ship foundered at Holyhead.

He later returned to South America. Prompted by the need to treat an eye condition contracted during his African travels, Lyon sought to return to England but died on the voyage home on 8 October 1832.

Despite such accomplishments, Lyon is curiously understudied. He does not figure in guides to world exploration, although his peers and contemporaries such as George Back, William Edward Parry, John Barrow, John Franklin, John Ross, James Clark Ross all do, nor does he feature in a recent discussion of the narratives of Arctic exploration.[11] Yet, as Fleming shows, Lyon was one of 'Barrow's Boys', that significant group of African and Polar explorers under the direction of the influential John Barrow, Second Secretary to the Admiralty and the key coordinator of British voyages of exploration in the first 30 years of the nineteenth century.[12] Fleming portrays Lyon as a self-aggrandizing extrovert – as, in truth, he was: Lyon once described his main interests in life as 'balls, riding, dining & making a fool of myself'.[13]

Lyon merits our attention in relation to the above concerns because we have printed narratives of three different landscapes and environments – north African, Arctic, central American – a published private journal undertaken whilst on Parry's second voyage; correspondence between author and publisher; and correspondence from Lyon to Admiral Lord Bayntun, a life-long patron, in which Lyon writes from the Arctic and from Mexico of his experiences there. In chronological order (and in short-title only), the published narratives arising from Lyon's travels were four: *Narrative of Travels in Northern Africa in the Years 1818–19 and 20* (1821); a *Private Journal during the Recent Voyage of Discovery under Captain Parry* (1824); the *Brief Narrative of an Unsuccessful Attempt to Reach Repulse Bay, through Sir Thomas Rowe's 'Welcome', in his Majesty's ship Griper, in the year MDCCCXXIV* (1825) and the *Journal of a Residence and Tour in the Republic of Mexico* (1828). Each was published by the London-based firm John Murray. Others have shown how this publisher's archive can illuminate the production of narratives of Arctic exploration not as the straightforward printing of an author's manuscript – not, then, *en route* writing as a simple correspondence between different geographies and their textual rendition – but, rather, route writing as a process of mediation and practice, 'in the field' and 'back home'.[14]

Several broader questions thus become crucial. What are the relationships between narrative and practice? What are the material practices of route taking and route writing and the route into print? Discussing the relationships between narrative and practice in the Arctic, Bravo and Sörlin argue, with reference to the field sciences, that 'for scientific travellers in the field to work successfully – whether geologists, ethnologists or missionaries – they must also carry with them the narratives that lend their field practices plausibility'.[15] As they note, narrative is not a way of describing a practice: it is a practice in its own right. Since book historians, geographers and others have recognized that what they are examining in printed narratives is less fixity than it is compromise and inconstancy, the relationships between author and publisher, between manuscript and print, become important to interrogate.[16] What does *en route* writing understood as a process of mediation rather than of presumed correspondence between the world and our words reveal about the practices of route making in the field?[17] These are my concerns in examining Lyon's work and writings, chiefly here his African and Arctic narratives.

ROUTE MAKING, ROUTE WRITING AND THE ROUTE TO TRUTH

Lyon begins his *Narrative of Travels in Northern Africa* (1821) with a conceit characteristic of many authors of travel narratives:

> The situation of an author, when he presents himself to the scrutinizing observation of the public, must ever be one of the greatest doubt and anxiety; but as the following pages are intended only to detail facts in the plainest manner, without attempt at embellishment of any kind, it is hoped that they will not only meet with indulgence from the general reader, but, escape, without very severe comment, from the examination of the critic. All that can be said in their recommendation is, that they adhere strictly to the truth, and that not a single incident described by the author is in the slightest degree exaggerated; on the contrary, he has not only abridged but, in some instances, entirely omitted to mention circumstances which occurred to him, fearing either to excite doubt in the minds of his readers, or by too long details to trespass upon their patience.[18]

This declaration of veracity, the plainness of the style as appropriate to its purpose – 'facts in the plainest manner' and their correspondence to truth – and the modesty of the author in order not to test the credulity of readers is at one with what others have termed 'epistemological decorum' in the production of works of narrative and scientific fact, and part of conventionalized prefatory remarks in travel narratives.[19]

If his work was thus depicted as truth-full and trustworthy and self-redacted, reservations attach to its author in at least one respect. In travelling, Lyon and his companions disguised themselves so as to assure safe passage: 'Mr Ritchie [the British Consul] assumed the name of Yussuf el Ritchie, Belford was called Ali, Dupont Mourad and I was styled Said Ben Abdoullah Allah'.[20] What was common sense in one context was in another an explicit statement about the moral conduct of the British traveller being based in deceit: Lyon was not who or what he appeared to be. As Heffernan has shown of the near contemporary search for Timbuctoo and the success of the disguised French explorer, Rene Caillié, in contrast to the death, in military uniform, of the British officer Alexander Gordon Laing, to travel in disguise was judged by the British authorities as morally reprehensible: how could one trust the truth claims of a traveller who did not correspond with what he appeared to be?[21]

Drawn from Nature by F.G.Lyon. On Stone by D.Dighton.

A SAND WIND ON THE DESERT.

London Published by J. Murray Albemarle St. Feb.1.1821.
C.Hullmandel's Lithography.

Figure 6.1 The hardships of Saharan travel are clear in this scene – 'drawn from nature' as Lyon has it – from Lyon's 1821 African narrative
George F. Lyon, *Narrative of Travels in North Africa* (London, 1821), opposite p. 70. Reproduced with permission of the Trustees of the National Library of Scotland.

Four elements characterize Lyon's North African *Narrative of Travels*: a rather abrupt style of writing; attention to people and to landscape in which dual context he is interested in naming and classifying; his use of sketches, landscape scenes and ethnographic portraits in order to depict and site places and customs; and reference to thermometry: Lyon often began description of the day's travel with a note about temperature. He provides in his *Narrative of Travels* a 'Meteorological

Register' for the period May 1819 to 20 October 1819, at which point the record stops: 'From this time until the third of December Belford and I were confined to our beds'.[22] Bodily frailty was not the only impediment to the practice of regular thermometry:

I think it right to account for what might otherwise might be attributed to neglect, my having, in the latter part of the journey, omitted to notice the variations of the thermometer from the 28th of February to the present time. Not using my tent, I found much difficulty, in so large a Kafflé, in preventing stray camels or the slaves from treading on the Thermometer; and it was so frequently in danger of being broken, that I found no correct rate could be kept, and therefore gave up the attempt.[23]

Instruments other than the eye were important in a sea of sand without clear way markers (Figure 6.1). His attention to thermometry is important, however much Lyon was deterred by slave and camel in its operation. Thermometry was then coming into its own as a device of measurement and of scientific progress.[24] Natural philosophers in the late eighteenth century used ocean-temperature measurements of the Gulf Stream – 'thermometrical navigation' – in order to plot faster routes for Atlantic crossings.[25] Its application by Lyon in the Sahara was more than a quotidian practice which helped frame narratives of daily travel. Both man and instrument had to bear witness to qualities of reliability and working tolerance in different environmental conditions: such was, simply, to be expected of authoritative claims to truth. In being regularly kept and to a standard, scientific measurements would be useful as a calibrated record to readers and other later travellers.[26]

Lyon's 1824 *Private Journal* is similarly characterized by short factual statements. He makes extensive observations upon Esquimaux and their culture (Figure 6.2). His narrative is peppered by remarks concerned with a landscape, which depending upon the season, was either solid or liquid, by turn

Figure 6.2 A group of Eskimaux

G.F. Lyon in William E. Parry, *Journal of a Second Voyage for the Discovery of a North-West Passage from the Atlantic to the Pacific* (London, 1824), opposite p. 418. Reproduced with permission of the Trustees of the National Library of Scotland.

restricting or facilitating mobility of ship and author (Figure 6.3). Atmospheric conditions likewise hindered observation (a problem when navigation depended in part upon the stars or was hindered by wild swings in the compass needle due to variations in the earth's magnetic field): 'Astronomical observations, or good ideas respecting the lands, could not be obtained in consequences of the weather, which, with the kind of fatality that attended all my little excursions, was more than usually severe and foggy'.[27] Perhaps with a sense of fatalism, Lyon remarked simply at one point 'The dogs here broke my thermometer'.[28]

Lyon's *Private Journal* is of particular interest for being just that – private – and for its relationship to others' Arctic narratives. In the preface, Lyon makes clear that it was never intended to be published: 'it was written solely for the amusement of my own fireside, and without the most distant idea that it would ever see the light in any other shape than that of its original manuscript'.[29] Naval officers other than the cap-

Figure 6.3 The ships' route is here cut through the ice by use of saw, adze and surveying staff

G.F. Lyon in William E. Parry, *Journal of a Second Voyage for the Discovery of a North-West Passage from the Atlantic to the Pacific* (London, 1824), opposite p. 118. Reproduced with permission of the Trustees of the National Library of Scotland.

tain were permitted to keep journals only on condition that the work be submitted to the Admiralty at voyage's end: potential authors were restricted by governmental codes over secrecy and officers' conduct as well as over the content of their narratives. In this instance, it is precisely this restriction and its difference in content from official accounts that facilitated its publication:

> *Being sent with the other journals to the Admiralty, in obedience to Captain Parry's instructions, my friend Mr. Barrow, [...], advised me strongly to publish it, on account of the number of little anecdotes it contained relative to the habits and disposition of a people entirely separated from the rest of the world, and with whom we had for so great a length of time kept up an intimate and constant intercourse.*

He observed also, as an additional inducement, that Captain Parry, in his authentic and official account of the expedition, had not deemed it fit or necessary to enter into many of those minute and peculiar traits which are requisite for displaying the character of a strange people. Captain Parry's opinion on this subject agreeing with Mr. Barrow's, I could no longer hesitate; and therefore, after a few abbreviations, and the omission of some details of natural history, and of scientific observations, I sent the original manuscript to the printer.[30]

We have, then, a work not intended for publication being scrutinized by others and deemed worthy of publication because of its ethnographic emphasis. To accommodate others' justifications, the author excised the scientific content. His fellow navigator's work, Parry's, became the 'authentic and official' narrative by virtue of its attention to science, its author's military authority and his in-the-field guidance of others, including Lyon.[31] Yet the books were closely linked in other ways. At sea, Parry had directed Lyon what to sketch. 'I cannot deny myself the pleasure of bearing testimony to the obliging readiness with which Captain Lyon has always attended to my suggestions on this subject, as well as to the eagerness and assiduity with which he seized on every opportunity of exercising his pencil, which so monotonous and unpicturesque a voyage presented'.[32] Parry's narrative contains Lyon's work on the aesthetics of landscape and travel because he, Parry, had authority over Lyon as to what illustrations should accompany what sort of narrative account.

The Parry and Lyon voyage was instrumentally well-equipped: the *Fury* and the *Hecla* carried forty thermometers between them together with numerous other instruments of observation.[33] Thermometers were used on deck and to test the temperature of sea-water at various depths. The chronometers were used to fix 'with considerable accuracy' the geography of the coasts surveyed. Parry's account carries an 'Explanation of Technical Terms peculiar to

Figure 6.4 Immobility in route making in ships held fast in the ice afforded moments of recreation as well as of route writing: here, some of the crew play cricket

Note the line of blocks and ropes connecting the vessels for use when fog restricted visibility. G.F. Lyon in William E. Parry, *Journal of a Second Voyage for the Discovery of a North-West Passage from the Atlantic to the Pacific* (London, 1824), opposite title page. Reproduced with permission of the Trustees of the National Library of Scotland.

the navigation among ice'.[34] This was standard practice: instrumentation underscored navigation; safe navigation underscored narratives of it.

Such expeditionary practices of route making – fixing one's position, measuring land and sea, consulting with Eskimaux – were necessary for survival not just as a record of daily life. Narrating, or, more properly, writing by keeping a journal, private or otherwise, also countered boredom and the regularity of shipboard life, especially when held fast by the ice (Figure 6.4). Restricted as they were, Lyon and others would take walks across the frozen landscape, which, 'by habit, had possessed many points of interest'. As he further noted, 'Thus, although flat, and for above eight months entirely covered with snow, we had distinguished our walks by the high-sounding names of the promenade or causeway,

South-East Point, East Bay, Hills, Yackee Huts, Yackee Stone'. Their landscape was 'by habit' named and domesticated by virtue of the explorers' immobility. His larger route, the overall goal of the polar voyage, was also seasonally restricted, a fact which had consequences for his narrative. In May 1822, when thaw permitted a land expedition, Lyon's account of it is brief: 'As it is not my intention to give in my private journal an official report of an extremely uninteresting journey, I shall here observe, that I have rendered the account as short as possible'.[35] Here, Lyon restricted himself as, later, following Parry and Barrow's suggestions, he restricted his narrative by the removal of natural history and scientific observations.

Lyon had no such post-voyage authorial restrictions with respect to his *Brief Narrative of An Unsuccessful Attempt to Reach Repulse Bay* (1825). This was an official voyage prompted by Parry's second expedition, and included orders from the Admiralty that:

> *you should, not only yourself, but also those who accompany you, collect all such observations on the tides, currents, state of the ice, and other particulars, as may be useful to geography, and the navigation of the coast along which you are about to proceed, as well as to science in general; and you are also to collect as many specimens of natural history, in its various departments, as you shall have the means of carrying with you; and to make accurate drawings of such objects as may not, from their magnitude, be capable of being brought away.*[36]

Here, route-making was determined in advance by official orders. The route to truth was determined by direct observation, mediated instrumental examination and faithful reproduction through landscape sketches. Guided by Admiralty strictures upon his expedition's requirements, Lyon reverted in his *Brief Narrative* to a more factual prose, akin to his 1821 *Narrative of Travels*. He was again exercised by thermometry and drew upon past practice in its use and interpretation even when – precisely when – the weather got in the way:

Repeated observations of this kind have now brought to a certainty the assertion, that the approach to ice from an open sea, may be ascertained by the sudden changes of the thermometer; and acting from past experience, I caused the most attentive look-out to be kept, on observing it to fall suddenly on this morning. Yet this change first took place in a very thick fog, and we ran about ten miles before the ice was seen.[37]

Safe route taking could never be straightforwardly visual. Lyon's *Brief Narrative* contains many references to the regular practice of sounding – measuring the depth and geology of the sea floor. This measurement of an unseen landscape was crucial to safe passage – hidden shallows or rocks were

significant hazards to be mapped if possible – as was study of the tides, particularly tidal range (Figure 6.5). Route-making depended upon negotiating the invisible. With a tidal range as great as 23′ in some places, the relationship between what was sea and what land changed regularly and dramatically. That is why the Admiralty's instructions emphasized the need for information on tidal direction, range, strength and so on. This emphasis upon what William Whewell in the 1830s was to call 'tidology' was, with thermometry, part of the then emergent practices of oceanography and of the establishment of measurement-based professional practices in science.[38] Route making in the field was shaped in such ways. Route writing back home was shaped by institutional and authorial imperatives rather than by nature's agency.

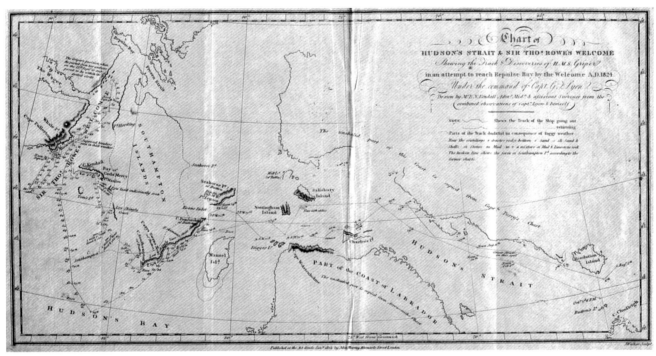

Figure 6.5 Route making required that its paper representation in map form be a guide to safe navigation

Note the uncertainty of many of the coastlines and the records of soundings and tides as the ship moves uncertainly past the island groups in the western end of Hudson's Strait. G.F. Lyon, *A Brief Narrative of an Unsuccessful Attempt to Reach Repulse Bay* (London, 1825), opposite title page. Reproduced with permission of the Trustees of the National Library of Scotland.

AUTHORIAL RESTRICTION, TEXTUAL REDACTION, NARRATIVE AND PRACTICE

The truth claims of travel narratives depend upon establishing the legitimacy of the author's experiences and an assumption, rooted in trust, that that experience corresponds in reliable ways to what is written. Authors make different specific appeals to truthfulness and employ different inscriptive and redactive practices to establish reliability and credibility. *En route* writing was about different routes to truth, different means to ensure correspondence between the landscape as encountered and the narrative as published.

We may, loosely, make a distinction between the practices which gave rise to Lyon's narratives, and the restrictions placed upon his narrative by matters of practice. In the first, we might place his bodily frailty, clumsy sled dogs, camels and slaves, regular but interrupted thermometry and other instrumental mediation, and the hazards of the landscape itself. For his Arctic travels especially, his routes were determined by the day-to-day management of a ship, often mobile but also often not. Route making there was hindered by limited light. In Africa, Lyon was dazzled by the light. In the second, we might identify Lyon's sense of the monotony of polar exploration, references to measurement as quotidian practices that guided his narratives' structure and encounters with landscapes which were, by turns, either covered by snow, obscured by fog or blown sand, or permanently hidden from view by the oceans and inferred only through instrumental mediation. His route taking and route making was determined in the field by practices of instrumentation and navigation and by degrees of indigenous knowledge. Route maps appeared as consequences of his work, not as guides to it. Lyon's narratives as routes to truth were influenced in their conception by the Admiralty's directives and in their completion by Barrow and Parry, despite the fact that, to take the 1824 *Private Journal*, they had been written only for the author's fireside amusement.

MacLaren concludes his discussion of the Arctic narratives of John Franklin and George Back with the caution that 'the findings in the case of one book or of one explorer are not necessarily pertinent to any other case'; 'Nor should the availability of publishers' correspondence with authors necessarily serve to undermine the status of the published text itself'.[39] Admitting the cogency of these remarks, it is clear that there is work to be done, text by text, route by route, on the relationship between author and publisher, manuscript and print, narrative and practice. Lyon's 'route into print' in 1824 depended upon others' views of its ethnographic merits and the author's own redactive practices. As an author whose printed works were altered *en route* to print by the publisher, Lyon is a weak example. As someone whose route writing was shaped in the field by particular practices and shaped by reviewers prior to publication through others' authorization, he is a very good example.

Attention in such ways to the nature and constitution of books reminds us that texts do not simply provide representations of objects in the field, but that they are themselves created through material practices of inscription mediated through social, political and economic institutions and agents.[40] The particularities of observation and inscription, redaction and mediation, and the questions of trust and correspondence upon which they depend are thus central to the production, communication and reception of geographical knowledge in print in the form of route writing. This is not to suggest that geographers and others engaged in the study of travel writing – writing about routes and about route making – cease to focus upon the representation of place or the practicalities of travel. It is to suggest that by examining the materialities of narrating and authorship it may be possible to understand more fully the epistemic bases to travel and observation and to recognize the practices of route making. From this, we may understand better the representational techniques which frame and condition those textual depictions central to our shared inter-disciplinary concerns in travel and route writing.

ENDNOTES

1   For general works, see Percy G. Adams, *Travelers and Travel Liars 1660–1800* (Berkeley: University of California Press, 1962); James S. Duncan and Derek J. Gregory (eds), *Writes of Passage: Reading Travel Writing* (London: Routledge, 1999); Julia Kuehn and Paul Smethurst (eds), *Travel Writing, Form, and Empire: The Poetics and Politics of Mobility* (London: Routledge, 2008); Justin Stagl, *A History of Curiosity: Theory of Travel, 1550–1800* (Chur: Harwood Academic Publishers, 1995); Jas Elsner and Joan-Pau Rubiés (eds), *Voyages and Visions: Towards a Cultural History of Travel* (London: Reaktion, 1999).

2   Rob Iliffe, 'Science and Voyages of Discovery', in Roy Porter (ed.), *The Cambridge History of Science. Volume 4: Eighteenth-Century Science* (Cambridge: Cambridge University Press, 2003), pp. 618–45; Harold Liebersohn, 'Scientific Ethnography and Travel', in Theodore M. Porter and Dorothy Ross (eds), *The Cambridge History of Science. Volume 7: The Modern Social Sciences* (Cambridge: Cambridge University Press, 2003), pp. 100–12.

3   Felix Driver, *Geography Militant: Cultures of Exploration and Empire* (Oxford, 2001); Miles Ogborn, '*Geographia's* Pen: Writing, Geography and the Arts of Commerce, 1660–1760', *Journal of Historical Geography* 30/2 (2004), pp. 294–315; Miles Ogborn, *Indian Ink: Script and Print in the Making of the English East India Company* (Chicago: University of Chicago Press, 2007); Robert J. Mayhew, 'British Geography's Republic of Letters: Mapping an Imagined Community, 1600–1800', *Journal of the History of Ideas* 65/2 (2004), pp. 251–76; Charles W.J. Withers, 'Writing in Geography's History: *Caledonia*, Networks of Correspondence and Geographical Knowledge in the Late Enlightenment', *Scottish Geographical Journal* 120/1 (2004), pp. 33–45; Charles W. J. Withers, 'Mapping the Niger, 1798–1832: Trust, Testimony and 'Ocular Demonstration' in the Late Enlightenment', *Imago Mundi* 56/2 (2004), pp. 170–93.

4   James Secord, *Victorian Sensation: The Extraordinary Publication, Reception, and Secret Authorship of* Vestiges of the Natural History of Creation (Chicago: University of Chicago Press, 2000); David Livingstone, 'Science, Text and Space: Thoughts on the Geography of Reading', *Transactions of the Institute of British Geographers* 30/4 (2005), pp. 391–401; Innes M. Keighren, 'Bringing Geography to the Book: Charting the Reception of *Influences of Geographic Environment*', *Transactions of the Institute of British Geographers* 31/4 (2006), pp. 525–40.

5   See, for example, Timothy Fulford, Deborah Lee and Peter J. Kitson, *Literature, Science and Exploration in the Romantic Era: Bodies of Knowledge* (Cambridge: Cambridge University Press, 2004); Nigel Leask, *Curiosity and The Aesthetics of Travel Writing* (Cambridge: Cambridge University Press, 2002); Kristi Siegel (ed.), *Issues in Travel Writing: Empire, Spectacle, and Displacement* (New York: Peter Lang, 2002); Frederic Regard (ed.), *British Narratives of Exploration: Case Studies of the Self and Other* (London: Pickering and Chatto, 2009).

6   Marie-Noëlle Bourguet, 'The Explorer', in Michel Vovelle (ed.), *Enlightenment Portraits* (Chicago: University of Chicago Press, 1997), pp. 257–315.

7   David Finkelstein, *The House of Blackwood: Author-Publisher Relations in the Victorian Era* (University Park: Pennsylvania State University Press, 2002).

8   Peter Lipton, 'The Epistemology of Testimony', *Studies in History and Philosophy of Science* 29/1 (1998), pp. 1–32; C.A.J. Coady, *Testimony: A Philosophical Study* (Oxford: Clarendon Press, 1992); Felipe Fernandez-Armesto, *Truth: A History and a Guide for the Perplexed* (London: Bantam Press, 1999).

9   Ian S. MacLaren, 'Exploration/Travel Literature and the Evolution of the Author', *International Journal of Canadian Studies* 5/1 (1992), pp. 39–68.

10 Ogborn, '*Geographia's* Pen', p. 296.

11 David Buisseret (ed.), *The Oxford Companion to World Exploration*, 2 vols (New York and Oxford: Oxford University Press, 2007); Janice Cavell, *Tracing the Connected Narrative: Arctic Exploration in British Print Culture, 1818–1860* (Toronto: University of Toronto Press, 2008).

12 Fergus Fleming, *Barrow's Boys: A Stirring Story of Daring, Fortitude and Outright Lunacy* (London: Granta Books, 2001), pp. 95, 105–6, 156, 162–70, 436.

13 Somerset Archives and Record Office, MS. DD/H1/D/553, Lyon to Bayntun, 23 December 1829.

14 Ian S. Maclaren, 'From Exploration to Publication: The Evolution of a 19th-Century Arctic Narrative', *Arctic* 47/1 (1994), pp. 43–53.

15 Michael Bravo and Sverker Sörlin, 'Narrative and Practice – an Introduction', in Michael Bravo and Sverker Sörlin (eds), *Narrating the Arctic: A Cultural History of Nordic Scientific Practices* (Canton: Science History Publications, 2002), pp. 3–32, quote on page 21.

16 On 'instabilities' in print, see David McKitterick, *Print, Manuscript and the Search for Order, 1450–1830* (Cambridge: Cambridge University Press, 2003), ch. 9.

17 This chapter is part of a AHRC-funded project addressing these questions: 'Correspondence: Exploration from Manuscript to Print, 1768–1848' managed by the University of Edinburgh (Institute of Geography and the Centre for the History of the Book) and the National Library of Scotland.

18 George F. Lyon, *Narrative of Travels in Northern Africa in the years 1818–19 and 20; Accompanied by Geographical Notices of Soudan and the course of the Niger* (London: John Murray, 1821), p. v.

19 Steven Shapin, *A Social History of Truth: Civility and Science in Seventeenth-Century England* (Chicago: University of Chicago Press, 1994), pp. 193–242.

20 Lyon, *Narrative of Travels in Northern Africa*, p. 7.

21 Michael Heffernan, '"A Dream as Frail as those of Ancient Time": The In-Credible Geographies of Timbuctoo', *Environment and Planning D: Society and Space* 19/2 (2001), pp. 203–225.

22 Lyon, *Narrative of Travels in Northern Africa*, p. 355.

23 Lyon, *Narrative of Travels in Northern Africa*, p. 343.

24 Hasok Chang, *Inventing Temperature: Measurement and Scientific Progress* (Oxford: Oxford University Press, 2004).

25 Charles W.J. Withers, 'Science at Sea: Charting the Gulf Stream in the Enlightenment', *Interdisciplinary Science Reviews* 31/1 (2006), pp. 58–76.

26 Randolph Cock, 'Scientific Servicemen in the Royal Navy and the Professionalisation of Science, 1816–1855', in David M. Knight and Matthew D. Eddy (eds), *Science and Beliefs: From Natural Philosophy to Natural Science, 1700–1900* (Aldershot: Ashgate, 2005), pp. 95–111.

27 George F. Lyon, *The Private Journal of Captain G.F. Lyon of HMS Hecla, During the Recent Voyage of Discovery under Captain Parry* (London: John Murray, 1824), p. 253.

28 Lyon, *Private Journal*, p. 420.

29 Lyon, *Private Journal*, p. v.

30 Lyon, *Private Journal*, p. vi.

31 William E. Parry, *Journal of a Second Voyage for the Discovery of a North-West Passage from the Atlantic to the Pacific; Performed in the Years 1821–22–23, in his Majesty's Ships Fury and Hecla, under the Orders of Captain William Edward Parry, R.N., F.R.S.* (London: John Murray, 1824).

32 Ibid., p. xiii.

33 Ibid., pp. x–xi.

34 Parry, *Journal of a Second Voyage*, ix–xx. Twenty-seven terms in all are given, of which twelve directly describe ice, by type, condition, location and impediment to navigation. Thus, for example, 'SLUDGE' – Ice of the consistence of thick honey, offering little impediment to a ship while in this state, but greatly favouring the formation of a 'bay-floe'.

35 Lyon, *Private Journal*, pp. 102–3, 208, 191.

36 George F. Lyon, *A Brief Narrative of An Unsuccessful Attempt to Reach Repulse Bay, through Sir Thomas*

Rowe's 'Welcome', in His Majesty's Ship Griper in the Year MDCCCXXIV (London: John Murray, 1825), p. xv.

37 Lyon, *Brief Narrative*, pp. 16–17.

38 Matthew Reidy, *Tides of History: Ocean Science and Her Majesty's Navy* (Chicago: University of Chicago Press, 2008).

39 Maclaren, 'From Exploration to Publication', pp. 51, 51–2.

40 Timothy Lenoir, 'Inscription Practices and Materialities of Communication', in Timothy Lenoir (ed.), *Inscribing Science: Scientific Texts and the Materiality of Communication* (Stanford: Stanford University Press, 1998), pp. 1–19.

Section II   The Route as Icon and Occurrence

This section traces moments in the development of the modern transportation landscape, looking at nineteenth-century canals and railways as well as early twentieth-century motorways. Starting by redefining the aesthetics of the sublime for a modern technological society, the chapters in this section scrutinize the relationship between the route and the landscape as an aesthetic, political and ethical issue.

7 Redefining the American Sublime, from Open Road to Interstate

David E. Nye

From the title, it might appear that this will be a story of cultural declension, one that begins with the inclusive optimism of Walt Whitman's open road and ends in an urban traffic jam. That tale would move from an open, unsettled landscape to sprawling cities, from walking with others to driving alone, from exploration to repetitive commuting. It might end with David Lynch's film 'The Straight Story' (1999), which can be seen as one man's attempt to recover the slower pace of Whitman's open road, a democratic space of chance encounters, camaraderie, and far more adventures per mile than the anonymous movement on the Interstate. This would be an easy and entertaining story to tell, but it would oversimplify.

Instead, consider the changing road as an expression of competing aesthetic practices. Some are more collective, others largely individualistic. Some emphasize speed and reliability, others privilege variety and flexibility. Every form of transportation, no matter how utilitarian it may seem, is implicitly part of an idealized landscape. It is not enough to look only at the vehicle and the pathway. Rather, each form of transportation has been used to define and construct a landscape, to embody a certain gaze. I will examine four examples of this process, focusing on the canal, the railway, the highway, and the Interstate.

ROAD AND CANAL

In 1855 Whitman published *Leaves of Grass*, a collection of poems that included his paean to the open road near the end of 'Song of Myself'.

> *I tramp a perpetual journey,*
> *My signs are a rain-proof coat and good shoes and a staff cut from the woods;*
> *No friend of mine takes his ease in my chair,*
> *I have no chair, nor church nor philosophy;*
> *I lead no man to a dinner table or library or exchange,*
> *But each man and each woman of you I lead upon a knoll,*
> *My left hand hooks you round the waist,*
> *My right hand points to landscapes of continents, and a plain public road.*[1]

In later editions of *Leaves of Grass* Whitman added another poem, 'Song of the Open Road', where he again evokes the pleasures of walking through various American landscapes. Likewise, Henry David Thoreau valued walking over any other means of transportation.[2] Whitman and Thoreau celebrated walking as a means to know not only the landscape but also one's self. However, during their lifetimes walking in practice

Figure 7.1 Five Locks of the Erie Canal at Lockport *(n.d.)*
Reproduced with permission of the Library of Congress, Washington, DC.

was beginning to give way to other forms of transport. In 1855 American western expansion was well underway, via the canal and railroad systems, which long remained in competition. Americans built many canals in the 1820s and 1830s, and then widened and deepened them in subsequent decades. As late as the 1880s more traffic moved on canals and rivers than on railroads.

In 1855, Americans could choose between multiple forms of transport, each with its own experience of travel: the road, the canal, or the railroad.[3] The roads reached the widest range of destinations, including wagon trails to California and Oregon. At that time, American roads were seldom paved, even with gravel. Most were eroded, marked by potholes, and unpleasant for travellers. Highways were often mired in mud, and stages were constantly getting stuck. Roads were frequently so bumpy that passengers were injured. An English traveller in 1839 found that one of the main roads into Philadelphia was 'composed of soft mud, nearly 18 inches deep, with alternate masses of unthawed clay and large stones'. He later had the crown of his head 'severely beaten against the top of the stage coach in the western regions' of New York.[4]

Canals were more comfortable and often faster. By the 1840s much of the region between the East Coast and the Mississippi was accessible by canal. Thousands of people, including the novelist Nathaniel Hawthorne, travelled the Erie Canal to the west, often to see Niagara Falls. Canal boats moved slowly, but were generally faster than wagons or other horse-drawn vehicles, because they could run day and night. Sleeping below deck, travellers awoke 25–30 miles closer to their destinations than when they went to bed. And in contrast to the continual jouncing and bumping of road travel, floating on the canal made it possible to sit tranquilly on deck, reading a book or contemplating the passing scenery. When the Erie Canal opened in 1825, early travellers thought their journey easy and comfortable compared to other options. They moved through the landscape at a walking pace, pulled by mules and horses. Passengers at times could get off the boat and walk along the towpath, and they often disembarked when boats passed through a lock. Indeed, towns sprang up wherever there were several locks close together, and passengers could eat or do a little shopping at sites such as Lockport, on the Erie Canal.

The pace of canal travel was unhurried and familiar, and the sense of space open and rather unfocused when compared to the railroad. Passengers often sat outside, with no obstructions to their view, and they were immersed in the landscape they passed through, which they could hear and smell as well as see. As Carol Sherrif notes, 'luxurious packet boats carried tourists on the "northern tour"', including many immigrants and foreign travellers.[5]

RAILROADS

Canal transportation was only fashionable for two decades at most. It was replaced by the faster railroads, which in America began to be built in 1830. Whitman may have celebrated the open road in his poetry, but when he wanted to visit the Rocky Mountains, Niagara Falls, or his brother in St. Louis he went by railroad. He felt that 'it was a happy thought to build the Hudson River railroad right along the shore, providing wonderful views en route from New York to Albany. Like most of his contemporaries, Whitman also liked looking at the railroad. He once noted, 'I see, hear, the locomotives and cars, rumbling, roaring, flaming, smoking. Constantly, away off there, night and day – less than a mile distant, and in full view by day. I like both sight and sound. Express trains thunder and lighten along; of freight trains, most of them very long, there cannot be less than a hundred a day'.[6] Americans had the world's most extensive system of railroads, which was built rapidly between 1830 and 1870, including the first transcontinental line, completed in 1869. The tracks spread from the Atlantic coast inland, and as early as 1840 the United States had twice as much track as all of Europe.[7] In subsequent decades, competition between mid-western cities ensured rapid expansion of the railways across the Great Plains and through the mountains.

Railway travel literally changed the way that people saw the landscape. In contrast to the canal boats, which floated low in the water to pass under the many bridges, passengers stepped up into railway passenger cars, which already were slightly elevated above the surrounding countryside. This height above the ground, combined with the speed of the train, made it difficult to look at anything nearby. John Stilgoe notes that even at 'thirty miles an hour, everything within the thirty or forty-foot mark appears blurred, unless the traveller is willing to swivel his head as the train passes. Increases in speed force the observer to look ever further from the car, and particularly east of the Mississippi River, such long views are rare'.[8] The railway journey erased the foreground, whose details disappeared from the traveller's experience. Only the larger and more distant panorama remained. This editing of the landscape, framed by the windows of railway cars, transformed the journey into the opportunity to see a limited number of sites, soon listed in guidebooks. The traveller was isolated from the passing scene, viewing it through plate glass, and could easily fall into a reverie, feeling that the train was stationary while the landscape rushed by. As Stilgoe notes, the people in the landscape glimpsed from the train 'struck passengers not as individuals but as a type'.[9] From a passenger's point of view their aesthetic function was to provide human scale. Railway travel inculcated a taste for the picturesque view, while eliminating a landscape's distinct sounds, textures, smells, and tastes, and preventing any direct contact between the traveller and local inhabitants.

What can one conclude about these three forms of travel in 1855? First, there are the continuities with the past. Both the canal and the railroad were social forms of transportation, in which the traveller constantly encountered strangers and typically fell into conversation. This was familiar from road travel in wagons and stages, and also from sea voyages. Indeed, storytelling among relative strangers can be considered a common feature of travel from at least the time of Chaucer's *Canterbury Tales.* Yet in other ways the canal and railroad broke fundamentally with previous forms of travel. Most obviously, the traveller became a passenger. Travellers had many active responsibilities. They were directly involved in moving themselves from place to place, either by walking, riding a horse, or, occasionally, getting out to push a stage or wagon stuck in the mud. Travellers also had to find food and accommodation along the way. Canals and railroads provided these things, simplifying the journey. Travellers frequently had to decide which road to take, and they needed to watch where they were going. But in the new transportation forms, passengers had little need to think about the route, which was unalterable. They could contemplate the scenery or even go to sleep. They also had less involvement with the landscape they passed through, which did not offer them physical challenges or impediments. On the canal boat a passenger could still smell and hear the passing scene, but on a train the outside world was apprehended only by the eye and seen through the rectangular framing of a car window. Most Americans

preferred the railroad to the canal, in effect declaring that separation was preferable to immersion and that speed was more important than the price of a journey. Indeed, a central feature of the railroad was its timetable, and the effort to be on time, a concept that meant less in traditional travel on roads, canals, or rivers.

Finally, both the canal and the railway delivered travellers to tourist sites *en masse,* notably at Niagara Falls. Both John Sears and Elizabeth McKinsey have noted how, after the coming of the railway in the 1840s, Niagara was converted into a series of distinct viewpoints, each of which charged admission.[10] Some of these viewing positions were man-made walkways, towers, and boats, while others were natural rock formations or islands in the river. Tourists were prepared to pay a good deal to see these vistas. One nineteenth-century tourist complained that it cost him eight dollars in fees to see Niagara Falls during two days, which is the equivalent of more than one hundred dollars today.[11] The Niagara experience consisted of paying admission to see carefully constructed views. The structure of this encounter is that which Dean MacCannell later described in *The Tourist.* The visitor is drawn to and recognizes a site by its iconographic representations and descriptions.[12] Off-site markers, such as illustrated railway calendars and brochures about Niagara Falls, aroused the tourist's desire to see, recognize and visually appropriate a site.

The aesthetic of this landscape was that of the sublime, as defined and illustrated in *American Technological Sublime.*[13] The railroad itself both celebrated and exemplified the technological sublime. The immensity of the system, the grand stations, the bridges leaping the rivers, the tracks scaling mountains, and the sheer speed of movement, all were triumphs over natural forces and limits. Yet at the same time passengers were deeply appreciative of natural landscapes. The railroad passenger's destination was often the Grand Canyon, Mammoth Cave, Niagara Falls, Yosemite, or Yellowstone. Indeed, the railroads at times gave free passes and free accommodation to painters who depicted the scenery along their lines. The Sante Fe railroad had a long-term arrangement with Thomas Moran, and reproduced his images of the Grand Canyon and the American Southwest on brochures and calendars for decades.[14]

HIGHWAY

The third transportation landscape, after the canal and the railroad, was that of the automobile, which began to emerge in the early twentieth century. In 1916 the railroad had reached its greatest dominance with 254,037 miles of track, or enough to crisscross the United States 80 times.[15] In addition, between 1890 and 1920 the United States built extensive streetcar systems. Some form of mass transit could reach most areas. One could travel from New York to Chicago on competing railroads, or even make the trip by trolley. In 1916 railroads carried 1 billion passengers, 98 per cent of all intercity traffic.[16] Yet precisely because railways were such a dominant form of transportation, reformers began to regulate them and occasionally to deny requests for rate increases. In 1906 they pressed through Congress stronger powers for the Interstate Commerce Commission, and from that time forward government controlled and approved railroad rates.[17]

Simultaneously, in American culture as a whole there was a widespread rejection of the railroad and a longing for travel that did not follow a single track or adhere to a strict timetable. This first emerged in the bicycle craze of the 1880s and 1890s, when thousands and then millions of 'wheelmen', and many women as well, took to the open road at a new pace, three times as fast as walking. With the development of pneumatic tires and the lower slung safety bicycle in the 1890s, the bicycle briefly became an important form of US transportation. A good deal of its appeal lay in the opportunity to explore new places that were too far to walk and inaccessible by rail. Just as importantly, the bicycle gave back to the traveller a direct, sensuous immersion in the countryside, along with the flexibility to go (and to stop) at any time.

The automobile had all of these advantages plus far greater speed and range. (Indeed, early popular fantasies included aerocabs and skies full of private vehicles, and as late as the 1930s Americans imagined that families would soon have private airplanes.[18]) There were only 8,000 passenger cars registered in the United States in 1900, and these were largely playthings for wealthy people who also commanded other forms of transportation. A railroad executive in 1900 could scarcely have imagined that the automobile and truck posed a serious threat. By 1913, there were 1.25 million registered automobiles (but only 68,000 trucks), or one for roughly every sixty-eight persons, which still suggested that motorized transport might be primarily for well-to-do people. Indeed, in Europe cars remained the prerogative of the upper class until after 1945. But in the United States, in 1913 Henry Ford introduced the assembly line, which in just ten years radically reduced the price of an automobile, from over $900 to less than $300. By 1930 Americans owned more than 26 million automobiles, or one for roughly every five persons.[19] The hegemony of the railroad was ending. From its peak in 1916, passenger railway traffic fell 30 percent to 700 million by 1930, and dropped to 450 million by 1940.[20]

The falling cost of the automobile alone does not explain the rapidity of its adoption. Americans preferred cars because they early associated them with everything that the railroad was not. Personal not corporate, never on a timetable, the automobile was understood as a vehicle of escape into adventures. The American West was particularly popular among early automobile tourists: 'the motorist's west stood for the same qualities associated with vagabondage and stagecoach days: independence, open space, simplicity, a more leisurely pace – all in direct contrast with the crowded, frenetic, compulsive East'.[21] Early accounts of automobile journeys emphasize exploration and encounters with the unexpected, in a countryside where the roads were poorly marked. One of the pleasures was the sense of roughing it away from the established tourist industry. Even flat tires and overheated engines, both then common, were seen as part of the automobiling experience.[22]

Adoption of the automobile at first meant that Americans sought to appreciate their landscapes in new ways, but soon it also meant that they created altogether new landscapes. Most obviously, automobile owners wanted paved roads. Asphalt roads had been introduced in the United States when some New York City streets were paved in the 1870s. Portland cement roads were also being built after 1894. These early efforts were promoted by sanitary reformers who wanted to reduce the mud and dust in cities, and by bicycle enthusiasts, who numbered over 1 million by 1900. In 1905, when there were still few automobiles, 161,000 miles of road had been paved, largely inside cities and between the most important centres. Stronger pressure to pave roads soon came from automobile manufacturers and owners, many of them farmers. Early federal laws funding road building emphasized rural areas, particularly postal routes, and discriminated against cities or heavily populated areas.[23] By 1925, there were 1.5 million miles of paved streets and roads, and federal funding had begun to include new intercity corridors as well as rural byways. During these same years, the railroad system gradually began to shrink, slowly until World War II, more rapidly after 1950. Freight remained important for bulk cargo, but passenger service disappeared on most lines, and what remained was abandoned by the private carriers and eventually consolidated into Amtrack.

Many early automobile journeys led to sublime natural sites, and sought to go 'off the main line'. One can see the automobility of 1900–1930 as an extension of the search for the natural sublime. But the developing automotive environment also fostered the new aesthetic of the commercial strip. First common on the roads leading to city centres, it soon grew up elsewhere, even on the roads into the Grand Canyon, Niagara Falls, and other natural sites. Along the new highways sprang up a myriad small businesses catering to passing motorists. These strips of services and entertainments created a new environment.[24] Most obviously, motorists needed car dealers, garages, tire stores, gas stations, and automotive supply stores in order to keep driving. The architecture of these establishments varied more widely than it does today, and included coffee shops that looked like hamburgers and 'colonial' gas stations. There were also mom and pop eateries, hamburger and hot dog stands, diners, restaurants, roadhouses (serving alcohol), miniature golf courses, tourist camps, tourist cabins, motels, markets, drive-in theatres, and drive-in banks. There were even drive-in funeral parlours, where the deceased could be seen from the car, as mourners gazed through a picture window at the body laid out in a casket. Often little or no zoning applied in these new regions, which grew chaotically and rapidly, and whose only organizing principle was the pursuit of profit. To attract drivers, these new businesses erected enormous signs that could be seen a mile away. They created a vernacular landscape of consumption, where bright lights and strange, eye-catching buildings vied with one another to catch the motorist's attention. There were cheese shops built to resemble an enormous round of Swiss cheese, shoe stores resembling a pair of boots, and diners that looked like a giant brown derby hat.[25] Moreover, 'Outdoor advertising, especially the billboard, followed the roads into the countryside and hawked all manner of products and services, including the new roadside businesses themselves'.[26] This visual cacophony dominated the road, and drivers seldom had an unobstructed view of a natural landscape. Indeed, a 1930s survey of the main road between Trenton and Newark New Jersey found that this 45-mile stretch of highway contained 300 gas stations, 472 billboards, and 442 businesses catering to the passers-by. To put this another way, there was a new business or billboard every 65 meters, and some used car lots, gas stations, or motels were at least that long.

It is easy to condemn this landscape as a disorderly, discordant, desert of banality. Kevin Lynch sounded the characteristic note of many urban planners: 'The commercial strip has many deficiencies – its noise, its confusion, its harsh climate, its monotony, its inhospitality to man on foot, its overwhelming ugliness. Strips are among the most 'polluted' man-made environments we have'.[27] New Deal planners, such as Benton MacKaye, sought to design more harmonious urban spaces that merged more seamlessly into their natural surroundings. Nevertheless, planned towns such as Norris were exceptions to the more general pattern.[28] Millions of Americans liked the commercial strip, and driving developed into a form of visual browsing. Like the Parisian *flaneur* on the boulevard, the driver on the strip was detached, wrapped in his own thoughts, aimlessly moving, enjoying an endless spectacle.[29] The roadside signs, businesses, and electrical displays, taken collectively, ceased to be merely advertisements. They defined an aesthetic that celebrated the man-made and the modern, and some architects would eventually champion it. In the 1960s Robert Venturi declared that the giant signs and colourful buildings were necessary symbols that could 'evoke the instant associations crucial for today's vast spaces, fast speeds, complex programs, and perhaps jaded senses which can respond only to bold stimuli'.[30] Venturi championed the new vernacular architecture in *Learning from Las Vegas*, which praised what others saw as garish and discordant.[31]

The commercial strip had emerged as a dynamic space of shopping, searching, promenade, flirtatious encounters, rivalries, and drag racing. Young people cruised the strip, looking for excitement. Visiting this commercial landscape became an end in itself, celebrated in popular songs and Hollywood films. For example, most of *American Graffiti* (1973) takes place either cruising along or at various locations on the commercial strip, including a parking lot, a drive-in restaurant, and other typical sites. The sound track consists of rock and roll songs heard from car radios, as four young men about to graduate from high school ride around town. There are guidebooks to what remains of this automotive landscape, such as *Roadside America: Gas. Food, Lodging*. It provides lavish illustrations and a breathlessly enthusiastic gloss on cruising, with much attention to sites along Route 66.[32] One of the most influential writers on landscape, J.B.

Jackson, also urged Americans to appreciate the vernacular roadside. He argued that there was much to appreciate in 'aspects of the landscape we have been taught to despise – the strip, the wide stretches of mechanized farming, the subdivision, the cheap resort'.[33]

The railroad had glided by scenery framed in large rectangular windows, and the difference between inside and outside was absolute. In contrast, the automobile could be stopped anywhere, and it implicitly abolished boundaries between the traveller and the surrounding scene. The new architecture did the same thing along the roadsides. Gas stations and other businesses reached out to motorists with capacious roofs and shaded parking areas, and they called to them with huge signs. This style of the 1950s and 1960s, variously called 'automotive moderne' or 'populuxe', had its origins in the art deco of 1920s and the streamlining and world's fair pavilions of the 1930s. Alan Hess has delineated this style,[34] whose main features included extensive use of glass windows that framed the street as a panorama, bold sign pylons that gave buildings a 'graphic presence on the street', and an architectural style that in contrast to streamlining emphasized acute angles, jagged lines, and cantilevered canopies and roofs. Finally, there was lavish neon and fluorescent lighting, making the new roadside architecture even more alluring and spectacular by night. In this new aesthetic, the car, the road, and the strip fused together into a single continuous landscape on the edges of American towns. Abolishing boundaries, 'The New Spaciousness allowed vision to flow from dashboard and windshield to the coffee shop table in one sweep'.[35] In such a space, the driver was integrated into a technological environment with little or no reference to nature.

If nineteenth-century landscape painters often depicted the railroad within a sublime landscape, in the middle twentieth century, photographers, painters, and architects responded to the highway billboards and the consumer culture of 'the strip'. The photographer Robert Frank traversed the new automotive space, capturing the emergence of the new aesthetic. A few of his images are about the road, and almost seem to be visualizations of Whitman's 'Song of the Open Road'. Yet the focus of his book, *The Americans*, often is on the popular culture encountered along the roadside. Indeed, Frank was most strongly drawn less to new shiny cantilevered environments of glass and neon than to the almost hand-made, artisanal world of the small diner or the little mom and pop businesses on the roadside. In contrast, Ed Ruscha made a systematic approach to more generic roadside architecture. He photographed 'Twenty-Six Gasoline Stations' during a 1500-mile trip that took 60 hours. His later work also tended to emphasize the standardized designs of commerce. Toward the end of the 1960s, Lee Freidlander further explored this space, but more subversively, taking many photographs from the pedestrian's point of view. His images continually reveal that the signs and principles of organization that help drivers to organize their perceptions have the unintended outcome of slicing up the pedestrian's world into incongruous parts.

INTERSTATE

The fourth transportation landscape, after the canal, railroad, and the commercial strip, began to emerge with the first automobile carriageways and parkways of the interwar years. The early American form of the limited access highway was not the same as the Italian or German roads built before World War II.[36] The new American highways were inspired in good part by Frederick Law Olmsted's carriageways in Boston, New York's Central Park, and Yosemite. American planners sought to build roads that would provide not merely transportation but recreation and enjoyment of nature. The sinuous curves and occasional steep hills of the suggestively named 'parkways' made them unsuitable for trucks. They were intended for automobiles only, and all horse-drawn traffic also was banned. By the 1920s, however, these aesthetically varied roads, designed for moderate speeds and enjoyment of the landscape, began to give way to more utilitarian turnpikes

and highways that were wider and straighter. In part, the newer roads were intended to provide a national network for military convoys, which had found the road system inadequate during World War I. Yet the most important pressure to build major intercity roads came from automobile and tire manufacturers and associations of car owners. Moreover, state governments provided a perpetual source of funding for more highways by dedicating gasoline taxes to this purpose alone.[37] The new highways were explicitly built for trucks as well as cars, and they were designed for high speed and safety. In Los Angeles, Phoenix, and other cities that grew primarily after 1930, the freeway became a central feature of urban life. These efforts to create a national network culminated with the passage of the Interstate Highway Act (1955).

By the late 1960s much of this Interstate system had been built and had become familiar, prompting Reyner Banham to argue that Los Angeles had four ecologies, one of which he named 'autopia'. He concluded 'the actual experience of driving on the freeways prints itself deeply on the conscious mind and unthinking reflexes'. It demanded 'a high level of attentiveness' and 'extreme concentration' which 'seems to bring on a state of heightened awareness that some locals find mystical'.[38] This state of reverie is, however, quite unlike the sublime. This sublime is induced by the shock of seeing the absolutely great, whether natural or technological, and it concentrates the mind on taking in a powerful object outside the self. In contrast, freeway driving is an intensive self-absorption in continual movement that leaves no time or spare mental energy for contemplation. Furthermore, the sublime is an encounter with something large and powerful, during which the person experiencing it, as both Kant and Burke explain, is in a position of relative safety. In contrast, the freeway driver is literally trying to avoid an accident and is always in a state of imminent danger. One false move, one careless driver, one blowout or one moment of inattention, can lead to maiming or death. Angelenos, Banham found, had to make an 'almost total surrender of personal freedom for most of the journey' during which they are immersed in 'an

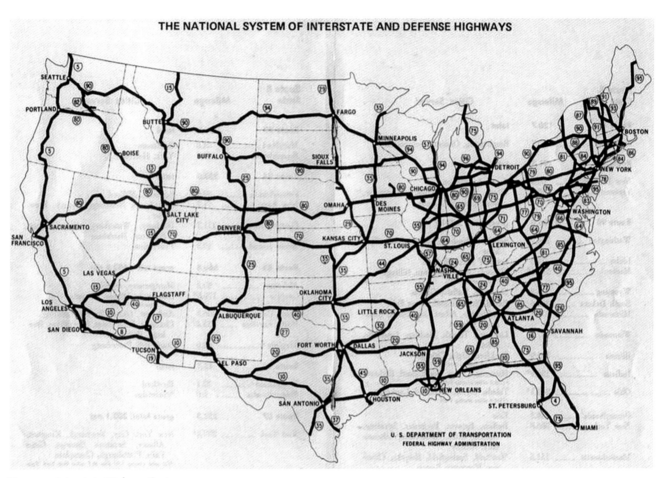

THE NATIONAL SYSTEM OF INTERSTATE AND DEFENSE HIGHWAYS

U. S. DEPARTMENT OF TRANSPORTATION
FEDERAL HIGHWAY ADMINISTRATION

Figure 7.2 Interstate Highway System, 1970
Reproduced with permission of the US Department of Transportation.

incredibly demanding man/machine system'.[39] The railroad or canal passenger could be inattentive or even go to sleep, but the driver must make a constant stream of decisions and manoeuvre within a shifting configuration of other vehicles.

Even before the advent of the Interstate system, the American relationship to landscape had been shifting away from the contemplative observation of a scene toward an interactive, high-speed encounter. This newer sensibility was manifested in a wide range of sports that became popular in the twentieth century: skin-diving, water-skiing, white-water

rafting, parachuting, car racing, downhill skiing, and many more. As J.B. Jackson has noted, each of these provided a sense of speed, a 'sense of danger or at least of uncertainty, producing a heightened alertness to surrounding conditions'. None of these new activities allowed 'much leisure for observing the more familiar features of the surroundings'. These sportsmen had a relationship to landscape quite unlike the railway passenger, who gazed out the window with no responsibility for controlling the movement or choosing the route. The interactive landscape of the driver was more abstract and 'seen at a rapid, sometimes even a terrifying pace'.[40] Since Jackson made these observations, other activities have emerged that confirm the trend, such as hand-gliding, snowmobiling, and roller-blading. People demand speed and immediacy, a maximum of experience in a minimum of time. Reaction times are short. Vision becomes central and the nuances of feeling through other senses are often ignored.

The automobile was a central part of this shift in awareness. The railroad had already accustomed people to passing through regions without being able to smell the plants or hear the sounds of the countryside, but the automobile gave each driver many choices, making it possible constantly to shift destinations, routes, and speeds. As with the new sports that emerged in the twentieth century, driving focused perception on an activity in the landscape, rather than the landscape itself. The new interstate roads often ignored the irregularities of local topography. Highway engineers blasted through hills, tunnelled through mountains, bridged valleys, and otherwise reshaped the land to conform to their specifications. They imposed uniformities on the Interstate, eliminating blind turns, steep gradients, and trees close to the highway. They designed uniform signage to be placed at regular intervals and at standard heights. Free enterprise was almost entirely eliminated from this landscape and restricted to a small number of service areas, each with a single gas station and food complex. The new Interstate system homogenized space so that the experience of driving through a new area resembled passing through a familiar place. On the Interstate, traffic lights were abolished. Constant movement was the norm, and stopping was prohibited outside the breakdown lane. This landscape encouraged speed and asked Americans to see landscape as generic background space, while focusing attention on the act of driving itself.

The new Interstates abolished the immediacy of the local strip, with its alluring distractions. The driver on the interstate was not a *flaneur* who could drop into myriad establishments on a whim. The limited access highway meant increased

and regular speed through a homogenous environment that did not draw attention to itself, and which was therefore not distracting. Indeed, the problem for the driver on an Interstate was not too much variety but sameness, potentially leading to inattention and daydreaming.

As the Interstate system grew, so did attacks on the older commercial strip. Pressure to ban both junk yards and billboards from the roadsides began as early as 1958, and culminated in the Highway Beautification Act of 1965.[41] But what was beautification to one class might be considered economic discrimination or even an unwelcome socialism by another. In 1958 a Senator from Oklahoma denounced attempts to ban billboards as attempts 'to deprive citizens of their vested rights' and compared the legislation to Russian communism and German Fascism.[42] While this rhetoric exaggerated, the new Interstate aesthetic was quite unlike that of the commercial strip. Interstate planners emptied the roadside of commercial signs and structures, in the same zoning impulse that led to suburbs that also excluded commerce or architectural variety. Fragments of the older strip survive, and new versions of the strip have also emerged along urban roads that give access to Interstates. But the architectural and visual counterpart to the Interstate became what Joel Garreau first called 'Edge City'.[43] Instead of a line of businesses each built according to the whims of its owners, the new mall complexes were far more homogeneous, self-contained, indoor environments, typically with few windows. Once inside, there was little to remind a visitor of the local topography or even the current weather. The apotheosis of the Edge City environment is the 'Mall of America' outside of Minneapolis, a vast block, with hundreds of shops, anchored at each corner by a department store.

The aesthetic of this fourth landscape is that of the consumer sublime, a term introduced in the final chapter of *American Technological Sublime* to describe some of the common features of the shopping mall, Disneyland, and the revamped Las Vegas.[44] At such sites, Americans shop for new sensations and new objects, with the proviso that they must be clearly labelled, neatly packaged, and easily appropriated. Las Vegas is an exemplary Interstate landscape, a fantasy world filled with generic symbols. Here the traveller recovers walking as the primary means of movement, albeit the walking is confined to a commercial, largely indoor space. In fact, many elderly people visit malls during off hours and quiet times to take their exercise in a climatized, crime-free environment. Teenagers go to the mall in somewhat the same spirit as their grandparents cruised the commercial strip. Shoppers enter the mall, often intent on finding a specific item, but after a few minutes of purposeful striding toward their goal become distracted by the intense, artificial environment that juxtaposes scenes from fairy-tales, history, advertising, novels, and movies. The commercial dreamscape of signifiers offers not the immensity of the sublime, but endless product innovation and variety. The vast cornucopia of goods and seemingly endless signifiers replace the immensity of the sublime landscape. For those with less cash, in more distant locations there are factory outlet stores in malls of their own. In the Interstate landscape, the man-made order seems completely to replace the natural.

EPILOGUE: THE DIGITAL LANDSCAPE

However, by the first decade of the twenty-first century, Americans were becoming tired of the automobile, just as they earlier tired of the railroad. Motorists are often so bored behind the wheel that they eat, make telephone calls, send text messages, and in a few cases even try to watch DVDs and drive at the same time. A psychologist studying such behaviour concluded that even 'the cell phone pulls you away from the physical environment. You really do tune out the world'.[45] Americans increasingly prefer a mediated environment, accessed through an iPhone, iPad, Blackberry, or laptop. These fuse together space and story, mind and matter, in a location that exists nowhere and everywhere.

The traveller is shifting away from the generic, repetitive landscape of the Interstate to the more compelling and endlessly varied digital landscape.

The traveller carries this digitized landscape, and looks at it rather than the people and places that are immediately present. The individual may be anywhere but is always virtually present at an e-mail address. The digital traveller never entirely leaves home, but remains tethered to Twitter, Facebook and email. By the same token, when not travelling the individual is never entirely immersed in the local either. Digital people are not rooted anywhere. Their journeys lead not to the natural sublime, nor to the technological sublime, nor to the peripatetic movement of the commercial strip, nor even to the homogeneity of the Interstate.

Americans apparently are becoming oblivious to physical space. Compared with the canal boat ride, the railroad journey weakened the sense of direct encounter with the passing scene, which was seen but not heard. Later, driving an automobile was used to escape the rigidity of the railroad, but at the cost of shifting attention away from the natural landscape toward the continuous challenge of driving. The commercial strip focused attention entirely on a humanly constructed world. Later still, the Interstate demanded an intense involvement in a man-machine complex while the natural world was transformed into a homogeneous and almost generic backdrop. Having eliminated the commercial strip, the Interstate landscape substituted the mall, where the consumer sublime completed the obliteration of natural surroundings and refocused attention on the cornucopia of goods. The consumer sublime at least remained focused on tangible things and actual sights, but the digital traveller and shopper of 2010 looks away from even its attenuated and stylized version of the physical world. While the driver must still look at the highway, passengers increasingly gaze at electronic devices or view films on screens embedded in the automobile's headrests. Children in particular seem to have lost interest in ordinary physical landscapes. The new endless road appears to be digital.

ENDNOTES

1 Walt Whitman, *Leaves of Grass: The First (1855) Edition*, Malcolm Cowley (ed.) (London: W. Scott, 1887), pp. 79–80.

2 Henry David Thoreau, 'Walking', in Charles W. Elliot (ed.) *Essays: English and American* (New York: Collier, 1914).

3 River traffic was also extensive, and important. See Louis C. Hunter, *Steamboats on Western Rivers* (Cambridge, Massachusetts: Harvard University Press, 1949).

4 George Combe, *Notes on the United States of North America (1841)* (New York: Arno Press, 1974), vol. I, pp. 298–99.

5 Carol Sheriff, *The Artificial River* (New York: Hill & Wang, 1997), p. 53.

6 Walt Whitman, *Specimen Days* (London, 1887), p. 204.

7 On railway competition between cities, see Zane L. Miller, *Urbanization of Modern America* (San Diego and New York: Harcourt College Publishers; Facsimile Edition, 1987), pp. 32–35. For more on railways see Alfred D. Chandler, *The Visible Hand* (Cambridge, Massachusetts: Belknap/Harvard University Press, 1977), pp. 122–144. On track in the United States and Europe, see Richard B. Morris, *Encyclopedia of American History* (New York: Harper & Row, 1970), p. 448.

8 John Stilgoe, *Metropolitan Corridor* (New Haven: Yale University Press, 1983), p. 250.

9 Ibid., p. 253.

10 Elizabeth McKinsey, *Niagara Falls, Icon of the American Sublime* (Cambridge: Cambridge University Press, 1985); John F. Sears, *Sacred Places, American Tourist Attractions in the Nineteenth Century* (Oxford: Oxford University Press, 1989).

11 Alexander Dow, *Anthology and Bibliography of Niagara Falls*, 2 vols (Albany: J.B. Lyons, 1921), pp. 1132–1133. Niagara developed first as a private site, but in the late nineteenth century it was taken over by the State of New York and the Canadian Government.

12 Dean MacCannell, *The Tourist: A New Theory of the Leisure Class* (New York: Schocken Books, 1976), pp. 118–125.

13 David E. Nye, *American Technological Sublime* (Cambridge, Massachusetts: MIT Press, 1994).

14 Ibid., pp. 75–76.

15 John F. Stover, *The Routledge Historical Atlas of the American Railroads* (London: Routledge, 1999), p. 54.

16 Ibid., pp. 52, 56.

17 Stephen B. Goddard, *Getting There: The Epic Struggle Between Road and Rail in the American Century* (Chicago: University of Chicago Press, 1994), pp. 186–193.

18 See Even Smith-Wergeland, 'Aerocabs and Skycar Cities: Utopian Landscapes of Mobility' in this volume.

19 Morris, *Encyclopedia*, p. 458.

20 Stover, *Routledge Atlas*, p. 56.

21 Warren James Belasco, *Americans on the Road: From Autocamp to Motel* (Baltimore: Johns Hopkins University Press, 1997), p. 27.

22 Ibid., pp. 23–24.

23 Owen D. Gutfreund, *Twentieth-Century Sprawl: Highways and the Reshaping of the American Landscape* (Oxford: Oxford University Press, 2004), pp. 26–27.

24 For a summary, see Larry R. Ford, *Cities and Buildings* (Baltimore: Johns Hopkins University Press, 1994), pp. 225–261.

25 On the strip, see Chester H. Liebs, *Main Street to Miracle Mile* (Baltimore: Johns Hopkins University Press, 1995).

26 Martin Melosi, 'The Automobile's Imprint on the Landscape', at http://www.autolife.umd.umich.edu/Environment/E_Overview/E_Overview6.htm.

27 Kevin Lynch, *City Sense and City Design* (Cambridge, Massachusetts: MIT Press, 1995), p. 579.

28 Elsewhere in this volume, see Christine Macy and Sarah Bonnemaison, 'The Concept of Flow in Regional Planning: Benton MacKaye's contribution to the Tennessee Valley Authority'.

29 Kathleen Hulser developed these ideas in a paper at the *American Studies Association* meeting in Kansas City in 1996.

30 Cited in Fran Schulze, 'Chaos in Architecture', *Art in America*, July–August (1970), p. 90.

31 Robert Venturi and Denise Scott Brown, *Learning from Las Vegas* (Cambridge, Massachusetts: MIT Press, 1977).

32 Michael Karl Witzel and Time Steil, *Roadside America: Gas, Food, Lodging* (St. Paul: Crestline, 2003).

33 Cited in Estelle Jussim and Elizabeth Lindquist-Cock, *Landscape as Photograph* (New Haven: Yale University Press, 1985), p. 106.

34 Alan Hess, 'Styling the Strip: Car and Roadside Design in the 1950s' in Martin Wachs and Margaret Crawford, *The Car and the City: The Automobile, The Built Environment and Daily Urban Life* (Ann Arbor: University of Michigan Press, 1992), pp. 167–179.

35 Ibid., p. 173.

36 See Christof Mauch and Thomas Zeller, *The World Beyond the Windshield: Roads and Landscapes in the United States and Europe* (Athens: Ohio University Press, 2008).

37 Gitfreund, *Twentieth-Century Sprawl*, pp. 22–29.

38 Reyner Banham, *Los Angeles: The Architecture of Four Ecologies* (New York: Harper & Row, 1971), pp. 214–215.

39 Ibid., p. 217.

40 John Brinckerhoff Jackson, 'The Abstract World of the Hot-Rodder', in Ervin H. and Margaret J. Zube (eds), *Changing Rural Landscapes* (Amherst: University of Massachusetts Press, 1977), pp. 146–147.

41 Carl A. Zimring, '"Neon, Junk, and Ruined Landscape", Competing Visions of America's Roadsides and the Highway Beautification Act of 1965', in Mauch and Zeller (eds), *The World*, pp. 94–107.

42 Cited in Zimring, p. 99.

43 Joel Garreau, *Edge City: Life on the New Frontier* (New York: Anchor, 1991).

44 Nye, *American Technological Sublime*, pp. 287–295.

45 Debbie Howlett, 'Americans Driving to Distraction as Multitasking on Road Rises', *USA Today*, 5 March 2004. See also Debra Galant, 'Driven to Distraction', *New York Times*, 20 July 2003.

8   The Man Who Loved Views:
C.A. Pihl and the Making of the Modern Landscape

Mari Hvattum

Figure 8.1 Viaduct at Selsbakk, *ca. 1863–64 (photo by C.A. Pihl)*
Reproduced with permission of the Norwegian Railway Museum, photo collection (JMF-1144).

Carl Abraham Pihl was a connoisseur of views. From the tower of his extravagant historicist villa in Christiania (now Oslo), he could look out over the roof tops of Homansbyen – a fashionable garden suburb which from the late 1850s on had become the home of Christiania's intellectual and political elite. Pihl gazed over the lush gardens surrounding the pavilion-like villas – an idyllic assemblage with a pleasing diversity of styles, forms and textures, composed almost in the style of an English park. As the art historian Ulf Hamran points out, Pihl had a particular sensibility for this picturesque urban landscape.[1] From his panopticon-like tower the city was like a garden – a picture – a panorama of style.

Pihl (1825–97) belonged firmly to the class of civil servants and academics who inhabited Homansbyen. Trained as an engineer in Sweden and England, he became Norway's first railway director, in charge of the state-operated Norwegian railway from 1865 until his death in 1897. His work for the railway had in fact started as early as 1851, when he oversaw the planning of the first Norwegian rail line between Christiania and Eidsvold, built by the British entrepreneur Robert Stephenson. In the decades that followed, Pihl was in charge of a multitude of railway lines, himself inventing a narrow gauge system that allowed railway construction even in very challenging terrain.[2] Personally supervising the construction work, Pihl travelled the Norwegian countryside up and down, subjecting it to radical transformation while at the same time documenting it with remarkable sensitivity. Accompanying him on his many trips was his collodium plate camera – an unusual contraption in Norway at the time, particularly applied *en plein air*.[3] Pihl was not put off neither by the rarity nor by the labour intensive process of glass plate photography, however. He dragged his camera, his stands and his plates down ravines and up cliffs, under bridges and along tracks. His photographic survey of the

Norwegian railway development chronicled a changing sense of landscape, documenting its transformation from an uncultivated wilderness into the beginnings of a technologically ordered territory. Photographing earthworks, tunnels, bridges and stations, Pihl subtly fused aesthetic conventions from landscape painting with a new-found fascination for technology.[4] In his carefully arranged images, the landscape is transformed into a scenery in a literal sense: a theatrical tableau on which a plot is played out. In Pihl's case, this plot revolved around the triad nature – nation – technology. Merging the dream of 'naturalness' with an ambition for technological progress, nineteenth-century nationalism prompted the making of a new kind of landscape in which nature and technology – the given and the made – were brought together in a precarious balance.[5]

PANORAMIC PERCEPTION

For Pihl, the railway was a means of opening the landscape not just for practical exploitation but for aesthetic enjoyment. In a passionate newspaper article promoting the building of a rail line through the Drammen region, he even refused to talk about practicalities at all: 'Don't be afraid that I will bore you with talk of rails and beams and batons [...]. [W]e should rather talk about nature's lovely scenery [...] the handsome Gravfos waterfall where the mighty water of the Drammen river presses itself through a crevice in the rock [...]. And the view from Sandmælen [...]; surely you have heard of it? I arrived there one evening, just as the sun was about to set. You won't believe how pretty it was'.[6] Pihl presented the railway as a vehicle for progress and aesthetic enhancement alike, making beauty and mobility available to all. With its 'unprecedented expansion of human communication' the railway would have 'quite an exceptional impact on cultural development and enlightenment' he argued.[7] For Pihl – a landscape photographer, hiking enthusiast and founding member of the Norwegian Tourist Association dedicated to making 'places of particular natural beauty accessible to the public' – enlightenment and natural beauty were two sides of the same coin.[8]

Railway travel has long been associated with views: with the pictorialization and aestheticization of nature. In his classic study *The Railway Journey*, cultural historian Wolfgang Schivelbusch introduced the term 'panoramic perception' to pinpoint just this: the way railway travel transformed the modern perception of the landscape from a place in which to act into a sight to behold. Paraphrasing Dolf Sternberger, Schivelbusch writes:

The railway transformed the world of lands and seas into a panorama that could be experienced. Not only did it join previously distant localities by eliminating all resistance, difference, and adventure from the journey: now that travelling had become so comfortable and common, it turned the travelers' eyes outward and offered them the opulent nourishment of ever changing images that where the only possible thing that could be experienced during the journey.[9]

Pihl's extraordinary landscape photography, produced over almost 40 years, could well be said to chronicle this change. In his newspaper petit he envisioned with relish the way the future railway traveller would light his cigar and immerse himself, from the comfort of his compartment, in the constantly changing scenery:

In front lies Modum's majestic massif, to the right, Tyristrand, already in darkness. Sigdal and Snarum's proud mountains are seen far away, still sunlit. Inserted into this lovely frame we saw the Tyrifjord and the Holsfjord with their blue waters stretching far, far into the distance. Under us lies Ringerike with its churches, its small forests and its pretty green fields over which solitary trees throw long, sharp shadows. Indeed; this is what we should talk about.[10]

Despite his aesthetic sensibility, however, Pihl's photographic oeuvre moves beyond a purely pictorial appreciation of the landscape. Both as an engineer and as a railway director he was personally involved in the immense physical labour of railway building, in the digging and drilling, the mining of tunnels, moving of earth mass, and clearing of forests. His photographs reflect this involvement. Pihl documented a reinvented landscape – a landscape transformed from inaccessibility to fecundity. It was a transformation that produced beautiful views, to be sure, but that also engendered a radically altered mode of interacting with the landscape. Rather than describing a neat evolution from active involvement to passive observation, Pihl's oeuvre testifies to a complex and convoluted relationship between the natural and the man made, in which technology and nature, aesthetics and utility were brought together in intricate ways.

THE LANDSCAPE AS SCENERY AND CONSTRUCTION: THREE PHOTOGRAPHS

Consider the following motif: A man – curiously out of place with a top hat and tail coat – is sitting with his back towards the viewer. He is overlooking a rather unassuming landscape – a slow, swollen river running to his right, pine clad hills rising gently in the distance. The man is perched on a log in a slightly awkward manner, his stiff posture seems to indicate that he was placed there more for compositional than recreational purposes. The man is observing what is obviously the focal point of the composition: a masonry bridge with four piers, crossing the river in the distance. The scene is composed with utmost care. Adopting the classical back-view from romantic landscape painting, Pihl's protagonist seems to be caught in an attitude of silent admiration for the scenery ahead. Like Caspar David Friedrich's *Der Wanderer über dem Nebelsmeer* (1817), the photograph presents a meditative landscape beheld by a solitary spectator. Even the man's dress has a striking resemblance to Friedrich's lonely wanderer.[11] Unlike

Friedrich's, however, Pihl's protagonist is not contemplating an untouched wilderness. If we look closely at the river bank with its battered bushes and its scarred plains, it becomes clear that this landscape has been subjected to heavy work, transformation, even destruction. The bridge – presumably the cause of all the effort – speaks to the same; a large-scale transformation of wilderness into a modern technological system. While borrowing conventions from romantic painting, Pihl – like mid nineteenth-century landscape photography in general – developed a different stance towards the landscape in which it was man's contribution to nature rather than nature itself that formed the focus. In this sense, Pihl's photography belongs to the same tradition as Andrew J. Russell's dramatic representation of the Union Pacific rail line across North-America, Roger Fenton's Crimean war photography, and Robert Howlett's coverage of British ship building.[12] Pihl's *Bridge over Glomma* particularly resembles William England's famous *Train Crossing the Niagara River* (1859), but unlike England's top hat wearing gentlemen for whom the railway bridge appears merely as a backdrop to

Figure 8.2 Bridge over Glomma at Kongsvinger, *ca. 1862–65 (photo by C.A. Pihl)*

Reproduced with permission of the Norwegian Railway Museum, photo collection (JMF-10363).

Figure 8.3 Tunnel construction at Spikkestad, *1870–71 (photo by C.A. Pihl)*
Reproduced with permission of the Norwegian Railway Museum, photo collection (JMF-00267).

their picnic, Pihl's *Rückenfigur* focuses intently on the bridge ahead.[13] It is man-made nature this elegant protagonist is admiring, a nature in which the machine, to paraphrase Leo Marx, has been firmly incorporated into the garden.[14]

At first sight, Pihl's wanderer testifies to Schivelbusch's panoramic perception. Nature is turned into a picture – a vista – a sight to behold. The bridge is inserted much like the pavilion in an English garden, a visual cue in a grand, emblematic narrative. And yet, the photograph captures not only a picture but a process: a dramatic and physical transformation of the landscape into a new kind of phenomenon. No longer the object of reverie, nature has

Figure 8.4 Construction work along the Kongsvinger line, *early 1860s (photo by C.A. Pihl)*
Reproduced with permission of the Norwegian Railway Museum, photo collection (JMF11306).

become subject of construction, alteration and improvement in a direct and dramatic way.

Nowhere does this drama get a more powerful expression than in the series of photographs showing earthworks and tunnelling along the Christiania-Drammen line. This was the region through which Pihl had argued so passionately for a railway in his 1857 newspaper article, presenting the natural beauty of the Drammen valley as a key argument for railway development. Invoking the educative power of beautiful scenery, he had portrayed the railway as a benign, almost invisible phenomenon, discretely opening up the landscape while itself receding unnoticeably into the background. Nothing, as Pihl himself knew very well, could be further from the truth. Railway building is a messy and violent activity, disrupting, transforming, and reshaping the landscape on a large scale. The earthwork photographs speak eloquently of the forces involved. Overwhelming in scale and with a strangely lunar texture, the huge earth mounds make the workers and surveyors look tiny, like ants. The slightly surreal sight of an engineer in a bowler hat peeking up from the mounds like a groundhog in February does little to eliminate the suspense. Far from the picturesque composition of the *Bridge over Glomma*, these images look almost unsettlingly modern, invoking a notion of the sublime which to twenty-first-century eyes seem more akin to François Lyotard than to Edmund Burke. It is not the sublimity of a grand, untouched nature that is being invoked here – the endless ocean or the overwhelming mountain. Rather, it is the sublimity of the man made, once it gets big enough, fast enough, or complex enough to take on the scale and incomprehensibility of nature. David Nye has aptly labelled it 'the technological sublime', referring to the particular fusion of technological power and natural grandeur that so thrilled the nineteenth century.[15] Pihl's railway photography, particularly his earthwork series, testifies to this notion in a powerful way.

Let us look at a third example. It is a photograph showing construction works along the Kongsvinger line, Norway's first state-run railway line, built between 1861 and –62. Again, Pihl adopted pictorial conventions from landscape painting. Two silhouetted figures with their backs half way towards the camera are placed on a rock to the left. The outcrop on which they are sitting – a rugged crag bearing marks of mining – falls steeply towards a gully at the centre of the photograph. Twisted trees in strangely distorted shapes line the opposite cliff, and between the two craggy rock faces appears the curvy line of a railway track. Flanked by a river on the left hand side, the tracks disappear perspectively into the horizon, their gentle curve transporting the viewer's eye into the distance. It is a curious photo, beautiful and haunting at the same time. The mysterious, silhouetted figures, the picturesque beauty of the rugged crags and twisted trees, could have come straight out of Friedrich. But whereas such elements in romantic painting usually mark untouched wilderness, here, being products of mining, they speak of something else all together. Just like the labourers toiling away on the tracks and the cultivated fields in the distance, they point, not to an untouched landscape but to a built landscape – created and used by man in a 'second creation', as Nye coins it in his book on the settling of America.[16]

The French historian of science, Antoine Picon, has pointed out the close relationship between modern engineering and the new notion of landscape emerging in the late eighteenth century. For the eighteenth-century engineer, the natural landscape was a malleable material to be formed and improved. '[T]he land, having been crossed, conquered and tamed, could be compared to a garden with the engineer as its foreman' Picon writes.[17] Some hundred years after the engineers from *Ecole des Ponts et Chaussées* inaugurated this new productive landscape, nineteenth-century railway engineers such as Pihl reaped its yield. It was a generation more appreciative of the natural landscape than their eighteenth-century predecessors, but who nevertheless maintained a distinct enlightenment ethos. For them too, productivity constituted beauty. And like their enlightenment counterparts, Pihl and his colleagues conceived their work as a process of cultivation – a charge to 'level the mountains, join up the seas, and render the uninhabited mountains fertile'.[18] While he certainly loved views, Pihl remained perhaps closer to Picon's enlightenment engineer than to Schivelbusch's nineteenth-century spectator. The beautiful view – like the beautiful garden – was the product of planning and labour: a landscape made, maintained, and represented by the engineer.[19]

THE TECHNOLOGICAL BEAUTIFUL

The images we have looked at so far open for contradictory readings. On the one hand they conform to the conventions of landscape painting – aestheticizing and pictorializing the landscape and turning it into a scenery. On the other hand, Pihl breaks with romantic landscape painting by showing us, not untouched wilderness, but the landscape as it has been shaped by man. In art historical terms he could perhaps be labelled a realist, not least in his unsentimental portrayal of railway workers. Pihl's workers were not the 'Sunday peasants' of romantic landscape painters such as Tidemand or Gude, but an unflinching presentation of labour and a detailed chronicle of technological transformation.[20] Pihl's photographs pictorialize nature yet they also reveal a notion of the modern landscape as, quite literally, a work in progress – a product of labour and ingenuity.

Pihl was fascinated with technology and his work may appropriately be discussed as example of the technological sublime. Yet his intriguing portrayal of railway lines and bridges testifies less to the overwhelming and the incomprehensible than to beauty, harmony and order. Rephrasing Nye's vocabulary slightly, one might perhaps speak of a 'technological beautiful' in Pihl's photography – a category encountered not only through the railway director's lens, but also in the cultural imagination of nineteenth-century Norway, in which this pragmatic-realist perspective had deep roots. For parallel to the panegyrics of national

Figure 8.5 Elverum Bridge, *ca. 1860–1870 (photo by C.A. Pihl)*
Reproduced with permission of the Norwegian Railway Museum, photo collection (JMF-11090).

romanticism ran a far more pragmatic enlightenment tradition, wary of the sentimental aestheticization of both 'folk' and 'land'. Poet and journalist Aasmund Olavsson Vinje, for instance – a friend and contemporary of Pihl – had launched an acerbic attack on the modern 'tourist gaze' in his travel memoir from 1861, arguing that the sentimental aestheticization of the wilderness was a sure sign of the alienation of modern man. 'No one should have to live here, except goats and Englishmen and art-lovers from the low-land' Vinje exclaimed spitefully when looking out over the jagged peaks of the Romsdal alps – a region as celebrated by tourists as it was hated by the local farmers, who found it barren and dangerous.[21] Vinje's solution was not to escape from the wild, however, but to improve it through communication and cultivation. Like Pihl, he was a member of the Norwegian Tourist Association, working tirelessly to make the mountains accessible for hikers. He wrote numerous poems about the Norwegian landscape and believed, like Pihl, in the educative power of beautiful views.[22] Both Vinje and Pihl represented that particular fusion of enlightenment pragmatism and romantic longing for nature so typical for the landscape discourse in mid nineteenth-century Norway. In this predominantly rural country where practically everyone originated from or lived on a farm, a landscape's usefulness and its beauty remained intimately connected. Neither the farmer boy Vinje nor the engineer Pihl ever stopped insisting on this link, even when they expanded 'use' to encompass the ephemeral business of aesthetic appreciation.

Carefully integrating modern technology into the natural landscape, Pihl's photographs presented not so much an awesome spectacle as the dream of a new harmony between technology and nature. His knowing adaptation of pictorial conventions from landscape painting speaks to the same ambition. Technology, here, is not presented as a break but as an enhancement of a wilderness waiting to be improved. Picon observes this ambition at work in eighteenth-century engineering: 'As far as [the engineers]

were concerned, joining up the countryside did not involve using any irremediable violence upon nature. The engineer's work did not so much destroy nature as reveal it'.[23] Nye, similarly, documents the way American settlers understood the clearing of the wilderness as a realization of nature's potentials; a project both natural and providential.[24] Pihl's photography conveys a similar idea of technology as something which completes and reveals the landscape. In the beautiful photograph of the *Elverum Bridge*, for instance, you see modern technology emerge from the agricultural landscape as its direct continuation. The traditional fence surrounding the field merges with the iron trusses of the bridge, and the stone piers form a reassuring backdrop to the hay stacks in front. Less interested in the sublime and awe-inspiring than in beauty and order, Pihl's landscape is a domesticated landscape in which modern technology is fitted seamlessly into the cultivated countryside.

An episode which aptly illustrates this domesticated notion of technology and landscape alike is Pihl's encounter with the Niagara Falls. An honorary guest of the Canadian railway, he journeyed to The United States and Canada in the summer of 1871. After weeks of travelling, Pihl reached what he obviously expected to be a highlight of the trip, the Niagara Falls, on 30 July. His reaction to this most celebrated of all sublime sights was, however, one of disappointment. 'The impression of the waterfall was gripping, but not what I had expected [...] The proud Clifton suspension bridge with its 1269' span, 300' above the river, was light and beautiful. The picturesque bridges across to Goat Island and 'Three sisters' – all in pleasant harmony – give to the place a decidedly attractive character' he noted in his diary.[25] To characterize Niagara in terms such as 'harmonious' and 'pleasant' may seem odd. This was after all a place, as Nye has demonstrated, invested with a cult-like status: a centre for the nineteenth-century veneration of both the natural and the technological sublime.[26] Pihl, however, was not a man given to rapture. More interested in the bridges than in the waterfalls (the former are described with exact measurements both in his letters and in his diary, the latter hardly mentioned), he found the scenery overrated but thought the bridges made the place interesting and pretty.[27] Technology, for him, was neither an eyesore nor a spectacle. It represented a harmonious addition to the landscape, an 'improvement' in the sense of eighteenth-century English landscape design. Pihl's technologically improved landscape is a landscape in which technology amends and perfects nature's deficiencies, fusing nature and culture into a beautiful and pleasant harmony.

RAILWAY PHOTOGRAPHY AND THE NINETEENTH-CENTURY TRANSPORTATION LANDSCAPE

Pihl's surprising characterization of the Niagara Falls provides a clue to the particular aesthetics at work in his photography. Rather than the technological sublime, Pihl cultivated a technological beautiful in which the awe-inspiring is downplayed for a notion of nature and technology as a harmonious *Gesamtkunstwerk*. The railway is presented as a seamless expansion of agricultural practices – a continuation, not a break, with Vinje's ideal fusion between utility and beauty. In his domesticated notion of the landscape, Pihl's work moves far beyond Schivelbusch's notion of a panoramic perception. While he certainly had an eye for the panoramic dimension of railway travel, Pihl conceived and portrayed the railway landscape primarily as a process of cultivation. And while he appreciated the aesthetic significance of technology, Pihl understood it not as a source of sublime spectacles but as an extension and continuation of the particular beauty belonging to the agricultural landscape. Photography was the perfect medium to capture this landscape. Opposed to romantic landscape painting with its emphasis on the wild and the old, mid nineteenth-century photography lent itself to the man-made and the new, perfectly capturing the aesthetic pragmatism of the farmer-cum-engineer. And if the eighteenth century saw nature as a potential garden waiting to be cultivated by the engineer, the nineteenth-century photographer made the engineer's garden subject of a rich aesthetic vocabulary. Pihl's railway photography offers an intriguing example of this process.

The young Pihl was dreaming of the beautiful views along the Drammen valley, envisioning the railway as a means towards enlightenment, both moral and aesthetic. As a mature railway engineer, Pihl actually shaped this territory; subjecting it to large-scale transformation and making it part of the new transportation landscape of the nineteenth century. Although this transformation was dramatic, it soon merged into the agricultural landscape as yet another step in the process of cultivation. Once the carefully calculated and laboriously constructed embankments were grown with grass and trees, their technological origin receded into the background. At that point, the tunnels would no longer seem like magnificent feats of engineering, but rather something that provided brief variations to the railway journey, a kind of visual syncopation brought on by the rhythmic shifts of darkness and light. Interestingly, the tunnel so dramatically portrayed as *work* in Pihl's photographs was described by contemporary tourist guides purely in terms of *views*:

What a marvellous surprise awaits the traveller on a sunny day as the train emerges out of the Røken tunnel. Suddenly, one finds oneself at the brink of an abyss from which one witnesses a landscape – the Lier Valley – so wonderfully lovely, so grand and rich, that the impression becomes unforgettable.[28]

Or as another guide book described it: 'Deep below one sees, like a post-card, the whole valley floor embedded between the Holsfjord and the Drammen fjord, strewn with villas and farmsteads, with meadows and forests, and all the picturesque effects offered by the Norwegian landscape'.[29] In a sense, then, Pihl's rhetorical equation of railways and views came true in the end. Once the landscape had recovered from the brutal construction work, the railway did indeed become a facilitator of views as much as an instrument of trade and transport. From the comfort of his compartment, the traveller could see a uniquely man made landscape – a second creation – all in 'pleasant harmony'.

ENDNOTES

1   Ulf Hamran, 'G.A. Bull. Homannsbyens arkitekt', *Fortidsminneforeningens årbok* (Oslo: Foreningen til fortidsminnesmerkers bevaring, 1959), p. 59.

2   Pihl himself wrote extensively on the narrow gauge system, see e.g. *The Railways of Norway: (More especially the Narrow-gauged Lines)* (Christiania, 1876). On the origins and history of the narrow gauge system in Norway see Roar Stenersen, 'Carl Abraham Pihl og Cap-sporet', *Jernbanemanden: Organ for norsk jernbaneforbund*, 1995, pp. 4–25. For a general introduction see Philip J.G. Ransom, *Narrow Gauge Steam: Its Origins and World-Wide Development* (Oxford: Yeovil, 1996).

3   Pihl learnt photography during his studies in England in the late 1840s, and became one of the first Norwegian photographers to apply the glass plate technique. See Vigdis Kraft, 'Carl Abraham Pihl: Jernbanebyggeren', *Norsk Jernbanemuseums Årbok* (Hamar: Norsk Jernbanemuseum, 1995). For a general survey of Norwegian photography, see Sigrid Lien and Peter Hansen, *Norsk fotohistorie: frå daguerreotypi til digitalisering* (Oslo: Samlaget, 2007). Surprisingly, this survey contains no reference to Pihl's extensive photographic practice, which remains little known.

4   The bulk of Pihl's photographic collection is held at the *Norwegian Rail Museum* and is searchable at http://www.norsk-jernbanemuseum.no/index.htm. The collection comprises around 250 images, many of them stereoscopic photographs. A few of Pihl's early photographs are presented in Harald Lund, Jonas Ekeberg and Bjørn Brekke (eds), *80 millioner bilder: norsk kulturhistorisk fotografi 1855–2005* (Oslo: Forl. Press, 2008).

5   The relationship between nationalism and landscape photography in nineteenth-century Europe is discussed by Jens Jäger, 'Picturing Nations: Landscape photography and national identity in Britain and Germany in the mid-nineteenth century', in J. Schwartz and J. Ryan (eds), *Picturing Place: Photography and the Geographical Imagination* (London: Tauris, 2003).

6   'Et Brev', anonymous letter in *Morgenbladet*, 31 October 1857. The railway historian Einar Østvedt attributes it to Carl Abraham Pihl in *De norske jernbaners historie* (Oslo: Norges statsbaner, 1954), vol. 1, p. 178.

7   Pihl, 'Et Brev'.

8   The Norwegian Tourist Association [Den Norske Turistforening/DNT], founded in 1868, was an important promoter of road and railway development. For a discussion on DNT and infrastructure, see Rune Slagstad, *Sporten, en idéhistorisk studie* (Oslo: Pax, 2008).

9   Wolfgang Schivelbusch, *The Railway Journey: The Industrialization of Time and Space in the 19th Century* (Berkeley and Los Angeles: University of California Press, 1986), p. 62.

10 Pihl, 'Et Brev'.

11 By the mid nineteenth century, the *Rückenfigur* was a wide spread convention in painting and photography alike, as discussed by Gernot Böhme in this volume. Pihl probably did now know of Caspar David Friedrich, who was generally little known at this time. See Werner Hoffmann, *Caspar David Friedrich* (London: Thames & Hudson, 2000), ch. 1.

12 Andrew Joseph Russell (1829–1902) started photographing in 1863 during the American Civil War. His celebrated portrayal of the Union Pacific Railway started in 1868 and was published in the popular folio *The Great West illustrated in a series of photographic views taken across the Continent taken along the line of the Union Pacific Railroad, west from Omaha, Nebraska* (New York:

Union Pacific Railroad Company, 1869). His images have many similarities to Pihl's, although Pihl started his photographic career about a decade before Russell. Roger Fenton (1819–1869) was chief photographer to *The Stereoscopic Magazine – a Gallery of Landscape, Scenery, Architecture, Antiques and Natural History*. Fenton was celebrated for his war photography and his portrayal of the English landscape and Pihl undoubtedly knew his work from publications and exhibitions in London. See Gordon Baldwin, Malcolm Daniel and Sarah Greenough, *All the Mighty World: The Photographs of Roger Fenton, 1852–1860* (New Haven: Yale University Press, 2004). Robert Howlett (1831–58), another celebrated British landscape photographer, was commissioned by the *Illustrated Times* in 1857 to document the construction of the world's largest steamship, the *SS Great Western* – an early example of industry as the subject of photography. Pihl lived in London at the time, and would undoubtedly have followed Howlett's photographic chronicle. For more on early industrial photography, see Helmut Gernsheim, *The Rise of Photography 1850–1810, the Age of Collodeon* (London: Thames and Hudson, 1987).

13 Like Fenton, William England was part of *The London Stereoscopic Society,* founded in 1854, and his stereoscopic photographs enjoyed tremendous popularity in Britain in the 1850s. See Ian Jeffrey, *An American Journey. The Photography of William England* (Munich: Prestel, 1999). Pihl, himself an early 'stereographer', was introduced to the technique during his stay in London in the 1850s. On the development of stereoscopic photography, see Gernsheim, *The Rise of Photography*, ch. 4.

14 Leo Marx, *The Machine in the Garden, Technology and the Pastoral Ideal in America* (Oxford: Oxford University Press, 2000).

15 David Nye, *American Technological Sublime* (Cambridge, Massachusetts: MIT Press, 1994).

16 David Nye, *America as Second Creation: Technology and Narratives of New Beginnings* (Cambridge, Massachusetts: MIT Press, 2004).

17 Antoine Picon, *French Architects and Engineers in the Age of Enlightenment* (Cambridge: Cambridge University Press, 1992), p. 217.

18 Extract from an eighteenth-century student dissertation from *Ecole des Pontes et Chaussées*, quoted in Picon, p. 224.

19 While Picon studies the representational strategies (mainly drawing) of eighteenth-century engineers, the role of photography for nineteenth-century engineering remains to be examined. Photography was taught in engineering schools both in England and France from the 1850s onwards, and the photographer-engineer was an established figure in the late nineteenth century (Gernsheim, *The Rise of Photography*, p. 19). A brilliant example is Georges Poulet (1848–1936), chief engineer of *La Companía Francesa Ferrocariles de Santa Fé* from 1889–1895. Poulet documented the railway building from Santa Fé to Tucumán in North West Argentina in beautiful cyanotypes, later published in *Georges Poulet: Aurora Argentina. Cyanotypien 1890–1894* (Munich: Galerie Daniel Blau, 2005). I am grateful to professor Micheline Nilsen at Indiana University South Bend for alerting me to Poulet's work.

20 Adolph Tidemand (1814–1876) and Hans Gude (1825–1903) were two of the most popular landscape painters of nineteenth-century Norway, not least for their collaborative painting *Brudeferden i Hardanger* [*Wedding Procession in Hardanger*] (1848), with its idealized presentation of Norwegian folk culture.

21 Aasmund Olavsson Vinje, *Ferdaminne frå Sumaren 1860* (Oslo: Samlaget, 1959), p. 211.

22 Ibid., p. 87.

23 Picon, *French Architects and Engineers,* p. 229.

24 Nye, *America as Second Creation*, especially ch. 7, 'Let us conquer Space'.

25 Pihl, diary entry, 30 July 1871. In Terje Colban Pihl and Anders Colban Følid (eds), *Jernbanebyggeren* (Oslo, 2000), p. 22.

26 Nye, *American Technological Sublime*, ch. 2.

27 Pihl, letter to his wife Kitty Pihl, 3 August 1871. In Pihl and Følid, *Jernbanebyggeren,* p. 24.

28 Christian Tønsberg, *Norge, Illustreret Reisehaandbog* (Christiania, 1874), pp. 39–42.

29 Christian Tønsberg, *Norge fremstillet i Tegninger* (Christiania, 1889), unpaginated.

9   Staging the Driving Experience: Parkways in Germany and the United States

Thomas Zeller

One of the elders of the discipline of landscape studies in the United States, John Brinckerhoff Jackson, aimed high when he envisioned an academic subdiscipline called 'odology' some time ago. Deriving the name for his field of study from the greek *hodos* (road, street), the ever-charismatic Jackson did not aim to study roads in and of themselves. Rather, odology was to be about the human shaping of roads and the ways that roads have shaped humans:

> Odology is the science or study of roads or journeys and, by extension, the study of streets and superhighways and trails and paths, how they are used, where they lead, and how they come into existence. Odology is part geography, part planning, and part engineering – engineering as in construction, and unhappily as in social engineering as well. That is why the discipline has a brilliant future.[1]

In the parlance of current science and technology studies, the co-evolution of human societies and their infrastructures were to be analysed. Jackson's call, however, has not been widely heeded among historians, as far as I can tell. Most historians of automobility have focused on cars rather than roads, with the former ones being attractive consumer goods which are privately owned and the latter ones being state-run infrastructures whose appeal is less obvious. Museums for the history of transportation and mobility mostly exhibit cars and rarely roads: To take but one example, the exhibition 'America on the Move' at the National Museum of American History in Washington, DC, is chock-a-block with trains and cars; the exception is one actual stretch of the myth-laden Route 66.[2]

One could conclude that Jackson's odology was merely an eccentric idea at an inopportune time. Few, if any, scholars have labelled their enterprises as odological; this is also true for the authors in this volume. But Jackson's call was heard: A growing number of academics in geography, American studies, and related disciplines have studied roads, roadscapes, and automotive landscapes, from intricately detailed accounts to sweeping overviews. Especially in the United States, the social and physical settings of motels, parking lots, fast-food restaurants, and gas stations, not to speak of the roads themselves, have been respectable research topics for at least a generation.[3]

These researchers have treated roads, their environs, and their usage as eminently social and cultural phenomena, thus echoing Jackson's basic belief approach that there is nothing inevitable about the relationship between humans and roads and in particular between roads and landscapes. Human activities and belief systems, economic calculations and political regimes have been responsible for changes in the physical construction, the design, and the perception of roads. Particularly, the view from the road has changed dramatically over the course of the twentieth century. By the end of this period, most fast roads were built, maintained, and perceived primarily as fast transportation corridors for the swift, ideally unimpeded movement of people and freight. In the early century, several roads in some countries were built as places themselves and avenues for scenic exploration. Generality and interchangeability characterize the former, while specificity and locality were the goals of the latter.[4] Understanding such changes requires putting them in a wide historical context. From a historian's point of view, it is also helpful to no longer treat landscapes and technologies as opposites, but to consider them as part of a larger spectrum.[5]

In this chapter, I am trying to understand landscapes that were specifically created to be seen by twentieth-century tourists in automobiles. In the United States, such roads were called parkways; in many European countries, mountain roads, lakeshore or seaside drives were built not simply to transport people in automobiles from one place to another, but to provide them with scenic views while they were travelling. Two roads stand out: the 750 kilometre-long Blue Ridge Parkway in Virginia and North Carolina, the longest American parkway, and its German counterpart, the *Deutsche Alpenstraße* [the German Alpine Road] extending 450 kilometres on the northern mountain crest of the Alps. Both roads were built as tourist parkways starting in the 1930s to stimulate traffic and open up relatively neglected tourist regions in the proximity of major population centres; and both presented particular versions of nature.

These roads opened a window on nature, made the natural environment accessible, and presented specific versions of nature. In the process, they acquainted millions of drivers and passengers with their countryside and rendered the surrounding scenery as easily consumable automotive landscapes. It was through these roads that many tourists learned to appreciate nature while they were on the road. Seeing and recognizing particular landscapes through the windshield became an important part of the nascent tourism industry. There was a decidedly national aspect of these roads as well, since the specific landscapes were advertised as carrying American or German qualities, or to be more

precise, the Southern variants of American and German culture. Both roads began as work-creation projects in the 1930s, albeit in the politically drastically different regimes of Germany's National Socialism and the New Deal of the Roosevelt Administration in the United States.

The landscapes in question here were not simply rural, agricultural landscapes. They were mountainous landscapes, whose peaks came to symbolize natural grandeur and whose roads were understood to mean a technology in synch with nature. But not every mountain and not every aspect of mountain life were incorporated into these ensembles. Therefore, it is interesting to understand which parts of the mountain landscapes have been chosen for presentation in touristic endeavours and how, in fact, the mountains themselves have become part of these automotive landscapes that parkways embody. The goal of these roads was more than simply carrying tourists to previously faraway and inaccessible places; according to the rhetoric of the designers, such roads would be able to reconcile the tensions between nature and technology through carefully designed landscapes. Demarcating the line between technology and nature is a historical creation; at the same time, the technology (in this case, road-building technology) that was proffered was quite distinct from the landscapes that it became a part of.

In the absence of regional or national highway systems, parkways became one of the most prominent forms of roadways in the United States from the end of World War I to the mid-1930s. These years also saw a dramatic rise of car ownership and usage in that country. While public transport in the form of trains and trolleys still flourished in these years, more and more Americans could afford to own automobiles and used them extensively, especially outside of the cities. By 1930, five Americans shared one car, statistically speaking. Many European countries did not match this level of motorization for another generation; in West Germany, this threshold was taken in 1965. Getting around in an automobile became increasingly popular in the United States during the interwar period. State administrations, which were in charge of building and maintaining roads, engaged in a veritable road-building frenzy. According to one historian, the mileage of paved roads increased from 387,000 in 1921 to 1,367,000 in 1940.[6] While thousands of miles of roads were adapted to the automotive age, parkways delivered more prominently on the promise of new construction of automotive highways. What is more, these roads were built for cars only, prohibiting trucks and buses by law, and promised to bring nature closer to drivers and passengers. Presenting attractive landscapes was not the only environmental boon that early automobiles were supposed to deliver; ridding American cities of the filth and muck of horse-drawn carriages was another. In the long run, however, the widespread use of automobiles created more problems than it solved.[7] As far as roads and landscapes were concerned, most observers agreed until the 1940s that parkways were an 'international model for the harmonious integration of engineering and landscape architecture', as the leading historian of parkways puts it.[8] Originating in urban public parks of the late nineteenth century, parkways soon became an integral part of American city planning. One of the country's most prominent landscape architect, Frederick Law Olmsted, coined the word 'parkway' in 1868 when designing Prospect Park in Brooklyn, New York. Primarily built for carriages, it had as few intersections as possible and was designed as the unity of roadbed and adjacent trees and shrubs, as a 'narrow, elongated park'. Neither trolleys nor commercial traffic were allowed. These principles were maintained as the parkway, a way through the park or from park to park, became increasingly used for automobiles. The historian Clay McShane argues that the prohibition of common carrier traffic on the parkways assured class segregation as well as the appropriate natural feel; social and environmental decisions were intertwined from the outset in the history of the parkway.[9] Increasingly, the design features of urban parkways were also utilized for extraurban parkways. Large rights-of-way were used to physically separate and visually screen the roadway from surrounding areas. The road itself was adapted to landform through a

Figure 9.1 The Bronx River Parkway in the New York City suburbs, showing the contrast between curvilinear parkway design and straight railway design

Reproduced with permission of the Westchester County Archives, Elmira, New York.

curvilinear alignment that preserved scenic features, such as streams and hills. Also, parkways introduced the idea of limited points of access, separate alignment for lanes running in opposite directions, and amenities such as roadside parks. One of the best-known extra-urban parkways was the Bronx River Parkway, leading to and from the affluent northern suburbs of New York, which was opened in 1922.

The commute on this parkway was naturalized to the extreme, with plants and trees shielding suburban development to passengers and drivers. Curvilinear road design predominated. The designers, among them Gilmore Clarke, a Cornell-trained landscape architect, extolled the ways in which both the natural and the social environment of the Bronx River Valley were restored through the planning and construction process: Instead of weedy and unkempt waterscapes, a pleasing and complete landscape was seen through the windshield. Poor immigrant families were displaced, as their dwellings gave way to the parkway and its wide right-of-way. The involvement of the eugenicist Madison Grant in the Bronx River Parkway project has led one recent observer to label the road as a quasi-eugenic project.[10]

Three lasting legacies arose from the Bronx River Parkway: the emergence of a vocabulary of beauty and accessible nature; the professional coalition of civil engineers and landscape architects; and the realization that property values alongside the parkways increased.[11] Between 1923 and 1933, New York's Westchester County spent over $80 million to complete a system of parks and parkways. Even more (in)famous is the work of Robert Moses as the chairman of the Long Island State Park Commission and subsequently as Commissioner of Parks for New York starting in 1934. With Gilmore Clarke as his chief landscape architect and a heavy-handed approach to public policy, Moses pushed for the construction of public works such as beaches, swimming pools, and urban parks, connected by hundreds of miles of parkways.[12]

It is worth noting that Moses and Clarke were only the popularizers, not the inventors of this idea of motorized access to nature.[13] Still, the Northeastern parkways (including the Taconic and Merritt Parkways) propelled these modern (in terms of their engineering standards) and naturalistic (in terms of their landscaping standards) roads to widespread prominence, both domestically and abroad.[14] Parkway historian Timothy Davis explains their popularity in paradoxical terms: 'While the parkway's practical benefits are readily documented, a more intangible reason for their widespread appeal was that by being both emphatically modern and resolutely traditional, they helped Americans negotiate the disjunctive experience of modernization during the tumultuous period between the two world wars'.[15] Several cities in the Northeast and the Midwest planned or built parkways in the inter-war period.

Also during the 1930s, the federal government sponsored the construction of the Mount Vernon Memorial Highway, leading from Washington to Mount Vernon and built to commemorate the 200th birthday of America's first president. The Bureau of Public Roads employed the same landscape architects who had worked in Westchester County and Bureau of Public Roads chief Thomas McDonald sought for 'as close an approach to nature as can be managed'.[16] Thus, parkways defined landscapes to the urban motorists and made them the vital ingredient of the simultaneously individualistic and pre-packaged weekend experience. Bringing nature closer to city dwellers was celebrated as a democratic achievement and thus a token of Americanness. In a historically rare coalition of professional groups, landscape architects and civil engineers presented parkways as a progressive means to egalitarian consumerism that would mend the rupture between country and city.[17]

The most prominent project of the time, however, was the Blue Ridge Parkway spanning 750 km in the southeastern Appalachians, introducing drivers to breathtaking views and pastoral ideals. As its most recent historian, Anne Mitchell Whisnant, notes, this parkway differs from the others not only in length. Unlike the Northeastern parkways, this federally funded road was not intended to ease commuting or make it more attractive.[18] Rather than serving the cultural and perambulating needs of relatively affluent urban and suburban dwellers, the Blue Ridge Parkway was intended to generate and sustain car-based tourism to one of the poorest areas of the country and to remake the landscape in view of the parkway, encompassing its human inhabitants as well as its fauna and flora.

Promoting tourism in one's own country had been the goal of the 'See America First'-movement, which exhorted Americans to spend their tourism dollars at home rather than in Europe.[19] Exploring and consuming scenic landscapes was part and parcel of this movement: The natural beauty, especially of parts of the Western United States, was intended to compensate for the lack of historic buildings such as castles and urban architecture that Europe provided.[20] While local tourism boosters had promoted such touristic domesticity for decades, the National Park Service, founded in 1916, and its first director, Stephen Mather, lent the support of the national government to these plans. Mather fervently believed that roads to parks and within parks were the key to increasing visitor numbers in the parks, thus enabling more visitors to gain more and better experiences of the parks' landscapes, and to elevate domestic tourism. He argued that little had been done 'to enable the motorists to have the greater use of these playgrounds [National Parks, TZ] which they demand and deserve' and pushed for more roads to be built.[21] Rather than arriving by rail, tourists would increasingly travel to National Parks in their cars. Improved highways would bring more visitors to the parks, thus resulting in their greater popularity and more calls for expanding the park system. In 1922, Mather supported the plans for a national highway system spanning the entire United States, propelled by his observation that 'Travel is based on the enjoyment of scenery'.[22]

Thus, it is not surprising that the National Park Service threw its institutional weight behind the idea of a road connecting two recently established National Parks, Shenandoah in Virginia and Great Smoky Mountains in Tennessee and North Carolina. Secretary of the Interior Harold Ickes deemed the project to be worthwhile of federal work-relief funds, as it would put thousands of unemployed Americans to work during the Great Depression and leave a regionally and nationally beneficial touristic infrastructure.

Built, beginning in 1935, through the cooperative efforts of state highway departments in North Carolina and Virginia

and federal agencies of President Franklin D. Roosevelt's New Deal, and completed in 1987, the Blue Ridge Parkway is still managed by the National Park Service today, and the management is total. The landscape writer Alexander Wilson has called it the 'prototypical environment of instruction'.[23] The Parkway's landscape architect and first superintendent, Stanley Abbott, who had worked on the Westchester County Parkways, sought to build 'a museum of managed American countryside'. At stake, then, was to design a landscape that would make 'an attractive picture from the Parkway'.[24] Given their metropolitan aesthetic sensibilities, landscape architects did not consider many vernacular dwellings and many of the local farmlands to be attractive for future motorists. In cases where farmhouses stood in the way of the new road, residents were resettled and offered new housing, often provided by the federal Resettlement Administration in the valleys.[25] The former farms were erased from the site, native succession species grew in their place. Any visitor today will realize the degree to which the National Park Service aims to establish visual control of the road and of the 'viewscapes', as the area surveyed from the inside of the car is called by planners today. In Wilson's words, 'The planners encouraged split-rail fences, grazing cows or sheep, but not abandoned cars or, for that matter, weeds'.[26]

At least initially, soil conservation on local farms in sight of the parkway or on parkway lands rented out to locals was a priority for the landscape architects and administrators of the new road. Through newsletters and meetings, locals were to be educated in better ploughing techniques or crop rotation in order to avoid runoff and establish more attractive landscapes. While such policies were in line with other federal initiatives during the New Deal, they soon took a back seat to the goals of expanding and maintaining the extensive road.[27]

The landscape architects were keen to let native plant communities resurface on the roadside. They were also very careful in controlling the vistas. Signs of undesirable human activity, such as industrial activity in the valleys, were screened from view with fast-growing trees and shrubs. The curvilinear road has speed limits of 35 or 45 miles/h; while slowing down when following yet another bend in the road, motorists are able to see for many miles. The techniques of alternating concealment and presentation, which garden and landscape architects had used for centuries to great effect in parks, thus took on automotive meanings. Given that the road is situated on the ridge of the mountains for considerable stretches, the views orient car drivers and passengers towards the valleys down below and the series of mountain ranges so typical of the Appalachian scenery. Therefore, 'motorists feel that they are on top of the world', together with their car, and in total harmony with nature: 'This national public landscape is organized around the private car and the private consumption of landscape'.[28] The intimate act of aesthetic appreciation is, at the same time, deeply intertwined with state activity and state control of the land and the view.

Figure 9.2 shows the harmonious relationship between fauna, flora, and drivers that the parkway planners intended to create. The overall message of this picture could be described

Figure 9.2 This bucolic image of the Blue Ridge Parkway dates from 1953 and was heavily used in advertising the road nationally
Reproduced with permission of the Blue Ridge Parkway Headquarters, Asheville, North Carolina.

as one of harmony. It is an expression of pastoralism in the literal sense as the sheep and their grazing grounds dominate the left half of the picture. No less harmonious is the road winding its way through the landscape and demarcating it, with a lone car taking possession of it. Split-rail fences mark the boundaries between technology and nature, but only reinforce an environment seemingly in balance.

According to Whisnant, such images obfuscate more than they reveal. She examines the way in which the parkway, as a grand public development fought over on the local, state, and federal level, deproblematizes itself.[29] In fact, a multitude of conflicts has shaped the parkway on all levels, from planning to construction to maintenance. Whisnant's political history of the parkway successfully debunks some of the myths of the only other academic book-length study of the parkway, which portrayed a road mostly in harmony with its social and natural surroundings.[30]

Whisnant's analysis can be complemented with a closer look at the kind of landscapes portrayed in this image. The geographer Stephen Daniels has coined the term 'duplicity of landscape' for such ensembles: Landscape often masks social relations and conflicts through its smooth and often aesthetically pleasant appearance.[31] The parkway landscape, in this respect, is duplicitous. The conflicts surrounding its creation have been naturalized and are not perceptible for the casual tourist. But this is not all. The Blue Ridge Parkway's creation has engendered contradictory reactions on several levels.

The Park Service's early interpretive plans for the Parkway reinforced urban clichés about rural 'mountaineers'. As modern as the road and the automobiles were taken to be, the pre-parkway world was portrayed as one of primitive living with few modern tools. On another level, parts of the educated, urban elites were not at all enchanted by the idea of a highway running along ridgetops in remote parts of the country. Rather than opening up this part of the Southeastern United States for touristic development and the eyes of urban multitudes, it would be better to leave it in a 'wild' state, some wilderness advocates argued. One historian notes that the meeting leading to the foundation of the Wilderness Society, an environmental group that successfully lobbied for the legal protection of wilderness areas in the 1960s, took place along an Appalachian roadside in 1934 while the Blue Ridge Parkway and other motorways were being discussed. Roads and especially ridge roads, for these advocates, became anathema to their vision of wilderness, defined as a place where humans are only visitors.[32] One of the wilderness advocates, Benton MacKaye, was the national driving force behind the creation of an East Coast hiking trail from Maine to Georgia, the Appalachian Trail which parallels and intersects with the Blue Ridge Parkway. Less successfully, MacKaye championed valley roads over ridgetop roads since he deemed them less intrusive. The landscape architects in the employ of the National Park Service had argued for a road alternating between ridge top and valley locations in order to gain a greater variety of views. But they were overridden in a political routing battle between the states of Tennessee and North Carolina. The latter offered higher elevations, prompting Secretary Ickes to award it the southern section of the parkway.[33]

Finally, it is hard to ascertain how visitors experienced the lookout terraces, winding curves, and mountainous locations of the Blue Ridge Parkway. Numbers indicate success: The National Park Service today counts more visitors on the Parkway than any other National Park. Whether all of them absorb the Appalachian scenery as the designers intended, is another matter. The variegated patterns of concealment and showcasing that landscape architects have imprinted on the mountains are meant to delight and stimulate. For some observers, however, the mere thought of hundreds of miles of automotive solitude surrounded by trees evoked the spectre of ennui. A journalist writing for the *New York Times* in 1938 imagined a ride on the Blue Ridge Parkway, describing the scenery to be enjoyed for hundreds of miles. But even before the road was finished, the metropolitan paper warned: 'Perhaps, after 600 miles of driving without sight of

a truck, you are a little homesick for the ordinary nuisances of touring, a little fed up with mile after mile of empty wilderness'.[34] This turned out to be an astonishingly prescient commentary: While the number of visitors to the Blue Ridge Parkway soared after World War II, some motorists avoid the parkway entirely or only drive on it for short stretches before seeking refuge in the familiar rites of navigating parallel four-lane interstate highways with common carrier traffic.

The scene for scenic roads in Germany was set in the political environment of the National Socialist dictatorship, whose planning styles differed fundamentally from those in the United States, even though some of the outcomes might look similar.[35] The politics of landscape consumption in Germany was closely connected to the environmental politics of the regime. It may be surprising to some that the most violent regime in European history expended rhetorical and legal resources for conservation and landscaping the countryside. Historians are still debating to which degree the rhetorical attention translated into systematic action.[36] One fruitful field of research is the roads built under Nazi rule. Directly inspired by the US parkways, the Nazi regime decided to sponsor the *Deutsche Alpenstraße*, a road extending 450 kilometres on the northern mountain crest of the Alps. Like the Blue Ridge Parkway, it was envisioned as a tourist road by local elites, especially in locales not connected to the railway network. In the Alpine countries of Austria and Switzerland, a handful of mountain roads were built in the 1930s with the dual purposes of luring automobile tourists and asserting national difference through the display of natural symbols such as mountain peaks. These roads were built as to incorporate the highest degree possible of 'Austrian' and 'Swiss' values as reflected in the cultural modernity of the road and its vernacular landscape. Numbers of car ownership and the beauty of the natural landscape were both seen as specific national achievements, thus imbuing car-driven mobility with a sense of modernity resting on state-sponsored road building. The fact that civil engineers extolled these roads as landscape-friendly as well as opening up the features of the landscape only reinforced the idea of a conciliatory triad of roads, cars, and landscapes.[37]

Accounts in middle-class touring magazines testify to the popularity of these roads. Central to their attraction was the view that motorists could gain from their cars. The Alpine panorama, of being able to see summit after summit from an elevated vantage point, had been the privilege of hikers and, increasingly, tourists who took cable cars to the top of mountains. The Zugspitze, Germany's highest mountain peak, was accessible by subway train as early as 1930 (today, it boasts Germany's highest internet café). However, middle-class automotive tourists preferred *not* to be in the company of either arduous, perspiring hikers or to be part of a railway tour, not even in first class. The panoramic journey famously described by Wolfgang Schivelbusch, the nineteenth-century landscape vision characterized by a loss of the foreground, took on social meanings as inter-war motorists aimed to regain the autonomy of movement and vision.[38]

Increasingly, motorists preferred to be in a company of their own choosing and to drive their privately owned vehicles themselves. By the mid- and late 1920s and early 1930s, chauffeurs began to disappear from travelers' accounts. By taking the wheel, these almost exclusively male drivers reasserted control over their machines and the panorama. The rediscovery of the foreground after the 'panoramic journey' prescribed by the railway was a major element of automotive narratives and desires. In a travelogue published in a middle-class motoring journal, Dr Elsa Bienenfeld traversed the Furkapass, 'one of Europe's most daring roads', in 1929. On top of the Rhone glacier, she experienced a noon hour filled with 'the magic of the most modern romance: mixture of grandiose nature and the art of the machine'. By 3 pm, a car park of some 300 automobiles had assembled, which she likened to an opera premiere combined with an Enrico Caruso concert, given the way the participants had dressed up and displayed their cars. The ensemble of cars was directed by a lone Swiss policeman, who in her estimation was dressed like an Austrian general.[39]

It was exactly this hybrid of automobiles and landscapes that these motorists were after. For many car drivers, the landscape impression was not a peripheral sideshow, but the central goal of their trip. The relatively low numbers for car ownership in Germany enabled this kind of visionary practice. Soon after gaining power in 1933, the Nazi regime promised to raise the level of motorization, but also promised to maintain and expand the idea of a motorized access to Alpine landscapes. Hitler announced an extensive program to motorize Germany by providing inexpensive, mass-produced cars and a nation-wide network of roads. The former effort in the form of the *Volkswagen* failed dismally during the Nazi years. However, some 3,800 kilometres of autobahn were built in the country.[40] In addition, Hitler himself made the Alpenstrasse his pet project in the summer and fall of 1933. Work on the road began in 1934. Like the Blue Ridge Parkway, whose construction began a year later, it was supposed to connect one summit with the next and traverse and bridge the valleys. The engineering challenges were enormous, and the design of the road changed. Instead of building a road which nestled closely to the mountains, by 1938 civil engineers resorted to building a road more domineering, less curvy, and more predictable. The plans included stretches as high as 1,700 meters.[41] Costs skyrocketed, thus postponing the completion of the road.

Even though it was incomplete and has never been built according to its original plan, the Alpenstrasse was celebrated in guidebooks and left a lasting imprint on the Alpine landscape. Very much unlike its Appalachian sibling, it touched upon or connected century-old towns and cities, which competed with each other over access to the road and thus tourism income.[42] As a result, the Alpenstrasse was routed as a pathway from one tourist attraction to another, from Hitler's mountain retreat close to Berchtesgaden in the East to Schloss Neuschwanstein, King Ludwig II's neo-romantic castle, in the West. Guidebooks published under the Nazis stressed that the Nazis' push for consumerism and their supposed respect for nature had been the enabling

Figure 9.3 The German Alpine Road promised Alpine views; its users were invited, if not instructed, to leave their cars at designated parking areas and to go for a hike

Hans Schmithals, *Die Deutsche Alpenstraße* (Berlin, 1936). Private collection.

factors for the roads. Motorists were constantly reminded that the landscapes which they experienced were German and essentially so; their car trips were supposed to reaffirm their belonging to an ethnically understood collective whose cultural values were expressed in its landscapes. The chief engineer for the road, Fritz Todt, instructed the drivers on the Alpenstrasse to be 'quiet, considerate in conduct, and reverential toward the grandiose nature surrounding you'. He also admonished them to thank Hitler.[43] Guidebooks placed baroque churches right next to valleys and mountains, thus creating a seamless web of nature and culture. Tourists were encouraged to leave their cars and go for hikes in the Alps (Figure 9.3). This managed landscape was an exclusionary one: Germans classified as Jewish could no longer ride their cars legally after 1938.

Not the least because of the Nazi sponsorship, work on the road did not continue immediately after the war. Some 60 kilometres were added and the road contributed to the staggering rise of Alpine tourism after World War II by opening up views of the Alps to non-hikers. But whether providing easy automotive access to the mountains was a desirable goal, a question that had not entered the public arena during the Nazi years, was very much under debate by the 1960s. When the Bavarian state government planned to construct an unrealized section of the Alpenstrasse from Linderhof to Füssen at the road's western end over the Hochplatte mountain (2082 meters above sea level) in 1965, conservationists were up in arms.

The German Alpine Club (Deutscher Alpenverein), a group of hikers with tens of thousands of members, spoke up against the project 'not because we begrudge the car tourists an attractive road link, but because a nature reserve which is unique in Germany since it is untouched, extensive, and particular, can easily be endangered'. The alpinists pointed out that the road would lead through a nature reserve established just two years earlier. Instead, the goal should be to preserve this spot of nature in its current state.[44] The project was indeed cancelled. When mass motorization had reached Western Germany, some sectors of its society deemed its mission of opening up the landscape obsolete. The plans for the road were never fully realized, and it was not until 2002 that road signs completed the process of inscribing the road unto the landscape.

Both the Blue Ridge Parkway and the German Alpine Road succeeded in their most obvious goal to increase car tourism to the regions through which they traversed. Especially in the Alpine case, tourism managers are now busy containing rather than attracting cars, thus questioning whether the Alpenstrasse is an unmitigated success. But less obvious comparisons between these two roads might be more helpful.

Far from being a transient moment in the history of motorization, the technology of landscape appropriation from the roadway offers a unique insight into one of the ways Western societies have tried to solve the conundrum of nature and technology. By consuming landscapes, major elites in the first half of the twentieth century aimed at bringing nature back into the technologically advanced societies of the West. In retrospect, this might seem like the opening of a Pandora's Box of nature destruction. However, the claim of the designers that they had found the means to embellish nature must be taken seriously and analysed historically.

The parkway ideal was one of total management of the land in the hand of a powerful public agency. While the National Park Service's agenda for the Blue Ridge Parkway was more discreet than Nazi Germany's blunt nationalism, in both instances the power of the state was meant to be visible, albeit to differing degrees. Interestingly, the statism of the parkway approach of consuming landscapes corresponded with a static understanding of nature: The landscapes designed were ideologically charged moments frozen in time, which required even more careful management in order to avoid succession dynamics of the flora.

These roads were built with the goal of not only offering scenic views to non-local tourists (mostly middle-class urbanites were imagined as users when the roads were planned), but also to transform the landscapes they traversed, especially in the Appalachian case. The Blue Ridge Parkway was, at least initially, unpopular with many local residents; state force, eventually, forced them to abandon hunting, logging, farming with previous methods, and moonshining in and around the parkway's landscapes. Environmental degradation narratives were instrumental for increasing experts' control. In the case of the Alpenstrasse, the reformist impulse was marginal, but the local population was often happy to serve as props in a scenery-driven automotive tourism display. In both cases these roads and their transformationist environments drew metropole and periphery closer together, established the dominance of expert knowledge, and reoriented the local economy from farming to tourism for outsiders.

It is striking to observe the parallel desire to make presentable and present a rigidly ordered version of mountain scenery from both roads. In the Appalachian case, the parkway was to serve as a way for inhabitants of the busy East Coast corridor to escape the cities and mid-century America. Instead, motorists encountered a monitored, prescribed, and comprehensive frontier landscape of settlers in 'primitive' terrain. The Blue Ridge Parkway narrative tended to exclude Native Americans and American history after industrialization, thus perpetuating a commonplace story of retreat and re-creation. Recreation, of course, on these roads made them into icons of car consumerism.

In contrast, the Alpenstrasse's planners took a more sweeping historical view, placing the topographical and human factors of the landscape in full view of the driver. Church steeples and mountain peaks taken together symbolized a cultural landscape with a particular Nazi bent: In a decidedly millenarian mood, the Nazi regime placed itself at the pinnacle of history; it was to be the result of landscape history and human history. The roads were presented to an ethnically cleansed German populace as a gift from its regime, indeed a gift from Hitler himself who put his political weight behind the project. Tourism as a part of the Nazi consumer culture, in this regard, was still paternalistic.[45]

Thus, the Blue Ridge Parkway was more than simply 'one of man's greatest achievements during the 20th century' as the American Society of Landscape Architects declared somewhat grandiosely in 2001 when it gave its 'Classic Award' to the road.[46] Together with other scenic roads, it is a testament to the tendency of some twentieth-century societies to solve the perceived tension between technology and the environment by designing and building a medium ground in the form of a roadway which, in the end, could neither solve nor acerbate these tensions.

ENDNOTES

1   John Brinckerhoff Jackson, 'Roads Belong in the Landscape', in Jackson, *Discovering the Vernacular Landscape* (New Haven: Yale University Press, 1994), p. 191.

This material is based upon work supported by the National Science Foundation (USA) under Grant No. 0349857. Any opinions, findings, and conclusions or recommendations expressed in this material are those of the author and do not necessarily reflect the views of the National Science Foundation.

2   To be fair, the exhibit portrays the history of interstate highways in some detail. But the artefacts are mostly those that move. Janet F. Davidson and Michael S. Sweeney, *On the Move: Transportation and the American Story* (Washington, DC: National Geographic, 2003).

3   For useful summaries see Timothy Davis, 'Looking Down the Road: J.B. Jackson and the American Highway Landscape', in Chris Wilson and Paul Groth (eds), *Everyday America. Cultural Landscape Studies after J.B. Jackson* (Berkeley: University of California Press, 2003), pp. 62–80; and John A. Jakle and Keith A. Sculle, *Motoring: The Highway Experience in America* (Athens: University of Georgia Press, 2008).

4   Christof Mauch and Thomas Zeller, 'Introduction', in Christof Mauch and Thomas Zeller (eds), *The World Beyond the Windshield: Roads and Landscape in the United States and Europe* (Athens: Ohio University Press, 2008), pp. 1–13; Marc Desportes, *Paysages En Mouvement: Transports Et Perception De L'Espace, Xviiie-Xxe Siècle* (Paris: Gallimard, 2005).

5   David E. Nye (ed.), *Technologies of Landscape. From Reaping to Recycling* (Amherst: University of Massachusetts Press, 1999); Sara B. Pritchard and Thomas Zeller, 'The Nature of Industrialization', in Stephen Cutcliffe and Martin Reuss (eds), *The Illusory Boundary: Environment and Technology in History* (Charlottesville: University of Virginia Press, 2010), pp. 69–100; Thomas Lekan and Thomas Zeller, 'Cultural Landscapes', in Andrew Isenberg (ed.), *The Oxford Handbook of Environmental History* (Oxford, forthcoming).

6   Bruce E. Seely, *Building the American Highway System: Engineers as Policy Makers* (Philadelphia: Temple

University Press, 1987), p. 141. Also, see the chapter by David Nye in this volume.

7   For a useful summary of these issues, see Tom McCarthy, *Auto Mania: Cars, Consumers, and the Environment* (New Haven: Yale University Press, 2007).

8   Timothy Davis, 'The Rise and Decline of the American Parkway', in Mauch and Zeller (eds), *The World Beyond*, p. 35.

9   Clay McShane, *Down the Asphalt Path. The Automobile and the American City* (New York: Columbia University Press, 1994), pp. 31–40; Clay McShane, 'Urban Pathways: The Street and Highway, 1900–1940', in Joel A. Tarr and Gabriel Dupuy (eds), *Technology and the Rise of the Networked City in Europe and America* (Philadelphia: Temple University Press, 1988), pp. 67–87.

10 Randall Mason, *The Once and Future New York: Historic Preservation and the Modern City* (Minneapolis: University of Minnesota Press, 2009).

11 John Nolen and Henry V. Hubbard, *Parkways and Land Values*, vol. XI, Harvard City Planning Studies (Cambridge, Massachusetts: Harvard University Press, 1937). Of course, such an economic observation holds true for most urban parks.

12 Many readers will be familiar with Robert Caro's and subsequently Langdon Winner's indictment of Moses as an authoritarian planner and the design of the Moses parkways as racially exclusive. In particular, Caro had claimed that some parkway bridges were built with a lower clearance in order to exclude poorer Americans, especially African-Americans, travelling in buses from using the parkways. According to Woolgar, Cooper, and Joerges, this is factually incorrect. By the late 1990s, New York's public transport agency was running daily buses on these parkways, thus refuting Caro's and Winner's claims on the surface of the road. Robert A. Caro, *The Power Broker: Robert Moses and the Fall of New York* (New York: Knopf, 1974); Langdon A. Winner, 'Do Artifacts Have Politics?' *Daedalus* 109 (1980), pp. 121–136; Steve Woolgar and Geoff

Cooper, 'Do Artefacts Have Ambivalence? Moses' Bridges, Winner's Bridges and other Urban Legends in S&TS', *Social Studies of Science* 29 (1999), pp. 433–449; Bernward Joerges, 'Do Politics Have Artefacts?', *Social Studies of Science* 29 (1999), pp. 411–431.

13 For a recent reassessment, see Hilary Ballon and Kenneth T. Jackson (eds), *Robert Moses and the Modern City: The Transformation of New York* (New York: Norton, 2007).

14 Davis, 'Rise and Decline'; Bruce Radde, *The Merritt Parkway* (New Haven: Yale University Press, 1993).

15 Davis, 'Rise and Decline', p. 37.

16 Timothy F. Davis, 'Mount Vernon Memorial Highway: Changing Conceptions of An American Commemorative Landscape', in Joachim Wolschke-Bulmahn (ed.), *Places of Commemoration. Search for Identity and Landscape Design* (Washington, DC: Dumbarton Oaks, 2001), pp. 123–177.

17 For related visions, see the chapter by Christine Macy and Sarah Bonnemaison in this volume.

18 Anne Mitchell Whisnant, *Super-Scenic Motorway: A Blue Ridge Parkway History* (Chapel Hill: University of North Carolina Press, 2006); Anne Mitchell Whisnant, 'The Scenic is Political: Creating Nature and Cultural Landscapes along America's Blue Ridge Parkway', in Mauch and Zeller (eds), *The World Beyond the Windshield*, pp. 59–78.

19 Marguerite S. Shaffer, *See America First: Tourism and National Identity, 1880–1940* (Washington, DC: Smithsonian Books, 2001).

20 Alfred Runte, *National Parks: The American Experience* (Lanham: Taylor Trade Publishing, 2010), p. 21, p. 34.

21 Mather as quoted in Ethan Carr, *Wilderness by Design: Landscape Architecture and the National Park Service* (Lincoln: University of Nebraska Press, 1998), p. 146.

22 As quoted in Carr, *Wilderness*, p. 147.

23 Alexander Wilson, *The Culture of Nature. North American Landscape from Disney to the Exxon Valdez* (Cambridge,

Massachusetts and Oxford: Blackwell Publishers, 1992), p. 35.

24 Abbott as quoted in Wilson, *Culture*, p. 35.

25 More residents, however, were displaced for the construction of Shenandoah Park. Justin Reich, 'Re-creating the Wilderness: Shaping Narratives and Landscapes in Shenandoah National Park', *Environmental History* 6 (2001), pp. 95–117.

26 Wilson, *Culture*, p. 35.

27 For New Deal conservation projects, see Neil M. Maher, *Nature's New Deal. The Civilian Conservation Corps and the Roots of the American Environmental Movement* (Oxford: Oxford University Press, 2008).

28 Wilson, *Culture*, p. 36.

29 Whisnant, *Super-Scenic Motorway*.

30 Harley E. Jolley, *The Blue Ridge Parkway* (Knoxville: University of Tennessee Press, 1969). For an examination of park roads in the American West, see David Louter, *Windshield Wilderness: Cars, Roads, and Nature in Washington's National Parks* (Seattle: University of Washington Press, 2006).

31 Stephen Daniels, 'Marxism, Culture, and the Duplicity of Landscape', in Richard Peet and Nigel Thrift (eds), *New Models in Geography*, vol. 2 (London: Unwin Hyman, 1989), pp. 196–220. Also, see Denis Cosgrove, 'Landscape and *Landschaft*', *Bulletin of the German Historical Institute, Washington DC* 35 (Fall 2004), pp. 57–71.

32 Paul Sutter, *Driven Wild: How the Fight Against Automobiles Launched the Modern Wilderness Movement* (Seattle: University of Washington Press, 2002), p. 233. For some examples on the debate surrounding wilderness in American environmental history, see William Cronon (ed.), *Uncommon Ground: Rethinking the Human Place in Nature* (New York: Norton, 1996); Michael L. Lewis (ed.), *American Wilderness: A New History* (Oxford: Oxford University Press, 2007).

33 For the political history of the routing controversy, see Whisnant, *Super-Scenic Motorway*.

34 Joseph Taylor, 'Parkway to the Wilderness', *New York Times*, 19 June 1938.

35 For an interpretation stressing commonalities, see Wolfgang Schivelbusch, *Three New Deals: Reflections on Roosevelt's America, Mussolini's Italy, and Hitler's Germany, 1933–1939* (New York: Metropolitan Books, 2006).

36 Franz-Josef Brüggemeier, Mark Cioc, and Thomas Zeller, eds, *How Green Were the Nazis? Nature, Environment, and Nation in the Third Reich*, Series in Ecology and History (Athens: Ohio University Press, 2005).

37 Walter Zschokke, *Die Strasse in der vergessenen Landschaft. Der Sustenpass* (Zurich: gta, 1997); Georg Rigele, *Die Wiener Höhenstraße. Autos, Landschaft und Politik in den dreißiger Jahren* (Vienna: Turia & Kant, 1993); Georg Rigele, *Die Grossglockner-Hochalpenstraße: Zur Geschichte eines österreichischen Monuments* (Vienna: WUV-Universitätsverlag, 1998); Cord Pagenstecher, 'Die Automobilisierung des Blicks auf die Berge. Die Großglocknerstraße in Bildwerbung und Urlaubsalben', in Thomas Busset, Luigi Lorenzetti and Jon Mathieu (eds), *Tourisme et Changements Culturels. Tourismus und kultureller Wandel* (Zurich: Chronos, 2004), pp. 245–264.

38 Wolfgang Schivelbusch, *The Railway Journey. The Industrialization of Time and Space in the 19th Century* (Berkeley/Los Angeles: University of California Press, 1986).

39 Elsa Bienenfeld, 'Eine Fahrt zum Rhonegletscher', *Motor Tourist* 9 (1929), pp. 14–5.

40 For the highly contested landscaping aspects of the autobahn, see Thomas Zeller, *Driving Germany: The Landscape of the German Autobahn, 1930–1970*, Thomas Dunlap trans. (New York/Oxford: Berghahn Books, 2007).

41 A.[ugust] Michahelles, 'Die Deutsche Alpenstraße', in *VDI. Zeitschrift des Vereins Deutscher Ingenieure*, vol. 82, 37 (10 September 1938), pp. 1067–1071.

42 Another difference is that trucks have been allowed to use the road from the beginning.

43 Fritz Todt, 'Geleitwort', in Hans Schmithals unter Mitwirkung von Adolf Stois und Waldemar Wucher, *Die Deutsche Alpenstraße* (Berlin: Volk und Reich, 1936), p. 5.

44 Deutscher Alpenverein, Dr Hans Faber, to Christoph [sic!] Seebohm, Federal Secretary of Transportation, 18 March 1965, Federal Archives Koblenz B108/62301.

45 Cf. Shelley Baranowski, *Strength through Joy: Consumerism and Mass Tourism in the Third Reich* (Cambridge: Cambridge University Press, 2004).

46 http://www.asla.org/meetings/awards/awds01/blueridge.html.

10 The Concept of Flow in Regional Planning: Benton MacKaye's Contribution to the Tennessee Valley Authority

Christine Macy, Sarah Bonnemaison

This chapter explores regional planning concepts developed by the American conservationist Benton MacKaye, and their influence on the early Tennessee Valley Authority. Trained as a forester, MacKaye used his understanding of natural systems to theorize the processes underlying urbanization. In this sense, we could say he was an urban ecologist, seeking to understand the dynamic forces that shape and transform cities.

MacKaye saw cities as systems of flows, involving the movement of populations, raw materials, energy, and transportation systems. His writings, and his influence within the Regional Planning Association of America, led to his appointment for a brief period at the Tennessee Valley Authority (TVA). This was a federal dam-building program established in 1933 for flood control, power generation and erosion control in the Tennessee Valley that, almost immediately, developed into a regional planning program involving land reclamation, new towns, regional parks, and highways. While MacKaye's tenure with the TVA was short,[1] he was one of the few people able to theorize TVA's regional plan as a complete system, involving 'the flow of water', 'the flow of products', and 'the flow of population'.

THE VIEW FROM ABOVE

In 1921, MacKaye wrote an influential article, 'An Appalachian Trail', in which he argues for the value of wilderness areas as an antidote to an over-urbanized society. In the first section of this essay, he imagines a giant standing atop the Appalachian mountain chain, his head scraping the clouds. 'What would our giant see', asks MacKaye, 'as he strode along the length of this skyline from north to south'?

Starting out from Mt. Washington, the highest point in the northeast, his horizon takes in one of the original happy hunting grounds of America – a country of pointed firs extending from the lakes and rivers of northern Maine to those of the Adirondacks. Stepping across the Green Mountains and the Berkshires to the Catskills, he gets his first view of the crowded east – a chain of smoky bee-hive cities extending from Boston to Washington and containing a third of the population of the Appalachian drained area.

Bridging the Delaware and the Susquehanna on the picturesque Alleghany folds across Pennsylvania, he notes more smoky columns – the big plants between Scranton and Pittsburgh that get out the basic stuff of modern industry – iron and coal. In relieving contrast, he steps across the Potomac near Harpers Ferry and pushes through into the wooded wilderness of the southern Appalachians where he finds preserved much of the primal aspects of the days of Daniel Boone. Here he finds, over on the Monongehela side, the black coal of bituminous and the white coal of waterpower. He proceeds along the great divide of the upper Ohio and sees flowing to waste, sometimes in terrifying floods, waters capable of generating untold hydro-electric energy and of bringing navigation to many a lower stream.

He looks over the Natural Bridge and out across the battlefields around Appomattox. He finds himself finally in the midst of the great Carolina hardwood belt. Resting now on the top of Mt. Mitchell, highest point east of the Rockies, he counts up on his big long fingers the opportunities which yet await development along the skyline he has passed.

Striding across the Appalachian mountains in his mind's eye, gazing at the panorama unfolding before his imaginary giant, MacKaye turns his thoughts not outwards towards the view, but inwards, towards the meaning and purpose of what he sees. As David Nye remarks in his contribution to this collection, walking – for nature-lovers like Thoreau, Whitman, and now, MacKaye – was a means to know oneself.[2] And for MacKaye, his thoughts were on the future of American society. He was concerned that Americans were becoming urbanized and alienated from nature.

As a social progressive and a passionate conservationist, he was frustrated that arguments for social progress were usually couched in urban terms. Even worse, the conservation movement of the early twentieth century was championed by elites as a vital resource to hone the manly instincts of young men – in short, it was militaristic, chauvinistic, and ultimately imperialistic.[3] MacKaye's contribution to the conservation

movement was his conviction that the wilderness experience was valuable for all citizens, but particularly for 'industrial workers and housewives'. It went without saying then, that Americans did not see themselves as alienated from wilderness, but defined by their relationship to it – at least in terms of their national myths.[4]

Figure 10.1 Young man on Appalachian mountain
National Archives, Washington, DC.

MacKaye's vision for reconciling wilderness experience with progressive social ideas was to propose the preservation of large tracts of wilderness along the Appalachian mountain range – much like the vast National Parks of the west, but easily accessible to the urban populations of the Eastern seaboard. In classic progressive conservation mode, MacKaye saw the Appalachian Mountains as sites of untapped resources (i.e. the 'black coal of bituminous and the white coal of waterpower'). Making a connection between natural resources and employment opportunities, he outlined a more socially oriented course for conservation policy.[5] The heart of his proposal was a series of camps linked by a walking trail along the mountain ridge. He imagined these as co-operatively run campsites that would grow into small communities offering sanatorium stays, agriculture and manufactures – combining much-needed leisure opportunities with the possibility of employment. In its broadest ambition, MacKaye's essay proposes a vision for the redistribution of an urban population.

He described his combination of a mountain trail and recreational camps as a 'flank attack on the problem of social readjustment'[6] – that is, a socialist agenda wrapped in the sheep's skin of a conservation plan. The view from the mountain would provide a 'chance to catch a breath, to study the dynamic forces of nature and the possibility of shifting to them the burdens now carried on the backs of men. The reposeful study of these forces should provide a broad gauged enlightened approach to the problems of industry. Industry would come to be seen in its true perspective – as a means to life and not as an end in itself'.[7] MacKaye wrote this essay in 1921, as he began his involvement with the Regional Planning Association of America (RPAA). This group welcomed his orientation towards parks and wilderness for the fresh perspective it brought to their own discussions about the future of the city. For MacKaye, the RPAA spurred him to locate his thinking within a planning framework and connect it with the problems of metropolitan growth.[8]

FLOWS IN THE METROPOLIS

MacKaye's proposal captured the public imagination, as conservationists and hiking clubs embraced the idea of an Appalachian Trail. It was even taken up by enthusiastic promoters of an emerging automobile tourism, who proposed a 'Skyline Drive' along the entire range[9] – to his horror, since he saw these groups as boosters for a cancerous extension of the city. In his next book, *The New Exploration: a Philosophy of Regional Planning* (1928), MacKaye presents an image of the metropolis as a system of flows. Bringing the reader high up in a tower in Times Square, he sees

> streams of traffic, passing through Manhattan, streams which flow and mingle with all the great traffic and goods streams of the earth. [...] The metropolis [is the mouth] that receives the industrial flow [... and] its hinterland [is the] industrial watershed, the parts of the country where the traffic streams take rise, first in small trickles and runnels, in farms, ranches, mines, forests, and then broadening into the vast streams of raw materials which go, as food or as basic products, into the homes and workshops of the world.[10]

While early stages of metropolitan growth saw flows going from hinterland to the city, in the twentieth century, argued MacKaye, the flow is outward and even more dangerous. Motor highways are the channels through which urbanization extends into the countryside: 'the motor slum in the open country is today as massive a piece of defilement as the worst of the old-fashioned industrial slums'.[11] Inverting the conventional trope, he described 'modern western mechanized civilization' as 'wilderness', using not the conservationists' positive value for the word, but its older meaning as wasteland, an uninhabitable and neglected place:

the conquering of one wilderness has been the weaving of another [...] through the industrial revolution a maze of iron bands has now been spun around the earth; this forms the modern labyrinth of 'industrial civilization'. And the unraveling of this tangled web is the problem of our day. [...] Will the framework which the genius of man has woven become a terrestrial lacework for the integration of his own terrestrial powers, or will it become a tangled net in which he will be strangled?[12]

Figure 10.2 Tennessee Highway 71, 1938
National Archives, Washington, DC. Record Group 142, ARC 280482.

His solution was to use wilderness preserves to 'dam' metropolitan growth. In the same way, greenbelts flanking highways would function as 'levees', containing the stream of urbanization that would otherwise overrun local culture and wilderness. By the early 1930s, these opposing views of the parkway idea had solidified into two camps: Metropolitan boosters saw them as corridors to extend urban commerce into rural areas, and regionalists saw them as instruments to decentralize cities by strengthening rural economies. MacKaye's view of roads as levees to control and direct urbanization falls within the latter camp.[13] He put forth this idea in an article written with Lewis Mumford in 1930, 'Townless Highways for the Motorist'.[14]

FLOWS IN THE TVA

Early in 1934, MacKaye was invited to join the TVA as a regional planning consultant, with the specific charge of drafting a comprehensive plan for the Tennessee Valley region.[15] MacKaye was thrilled by the possibility of applying his 'townless highway' idea on a large scale.[16] The first test of this idea was in the development of Norris Dam and its associated town, parks and freeway, a complex conceived as an integrated system that was to serve as an example of regional planning under the New Deal. The plan was ambitious, involving the construction of new dams, resettlement of an agricultural population that formerly inhabited the riverbeds, land appropriations and the establishment of a greenbelt reserve, innovative forestry and farming practices that would reduce erosion, small-scale industry, adult education, and park and tourism development.

MacKaye's boss, Earle S. Draper of TVA's Land Planning and Housing Division, agreed that Norris Freeway was 'one of the major features of the regional development plan' which set the stage for 'the full expansion of the resources and opportunities to be developed in the Valley'.[17] Describing the

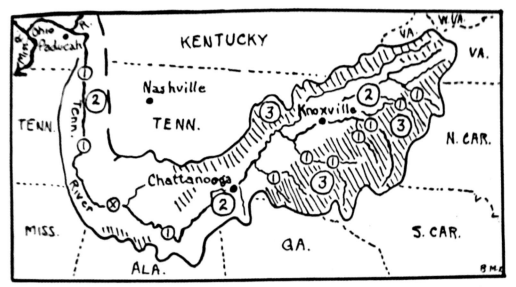

The Tennessee Valley plan for control and use of water flow. Figure 1 marks river regulation works, dams and reservoirs, 2, power lines to distribute current; 3, forest cover on slopes. X marks Muscle Shoals dam. Maps by the author

TENNESSEE–SEED OF A NATIONAL PLAN

BY BENTON MACKAYE

Figure 10.3 Benton MacKaye, TVA plans for the Tennessee Valley
Survey Graphic no. 22, 1933.

region around the first dam of Norris as knit by roads into a metropolitan cluster of rural trading and manufacturing centres, TVA planners saw the freeway as a means of accelerating the economic development of the region.[18] It would allow truck traffic to flow more easily, linking dispersed production to dispersed markets.[19] It would bring tourists to enjoy the scenery and purchase the craft commodities. Contained within arteries, traffic would circulate through the region and feed its system of exchanges.

'The problem of the Tennessee Valley', wrote MacKaye, 'comes down to "stream control" of two kinds: the stream of waters in the rivers, and the stream of development along the

highways. Each stream must be kept moving; each element must be finally distributed – one in terms of waterpower installed in the power lines, the other in terms of human beings settled in communities'.[20] For MacKaye, the TVA had to extend its 'sphere of influence' to include not only the flows of water and power, but of goods and people as well.[21]

In the frontispiece of an article published in *Survey Graphic* in 1933, MacKaye outlines the TVA's plan for 'control and use of water flow'.[22] He shows dams harnessing the waters of the foothills, power lines extending along the valley bed, and forest cover reaching up to the heights of the watershed. He extends the flow idea right up the Appalachian Mountain chain, indicating the direction of river flows with arrows: (1) indicates dams (2) power lines following the valleys, and (3) forest cover on the mountain slopes, roughly following his Appalachian Trail structure. Except this time, wilderness preserves are linked to urban areas by power lines, rather than roads and trails. In another diagram, MacKaye extended his thinking about flows to encompass population movements. The large dots are cities – the sources for the 'backflow' of population who would be drawn back to upland valleys by new communities provided with power, industry, agriculture,

and that would offer close contact with soul-restoring wilderness. (A) indicates the townless highways following the lower reaches of the mountain chain (B) highwayless towns, and (C) preserved wilderness areas.

Flow control was MacKaye's paramount concern during his tenure at the TVA. This married nicely with the Authority's primary purpose of harnessing the waters of the Tennessee River to energize and renew a region. And indeed, the imagery of flow extended through the design projects of the Authority. The dams, of course, transformed the power

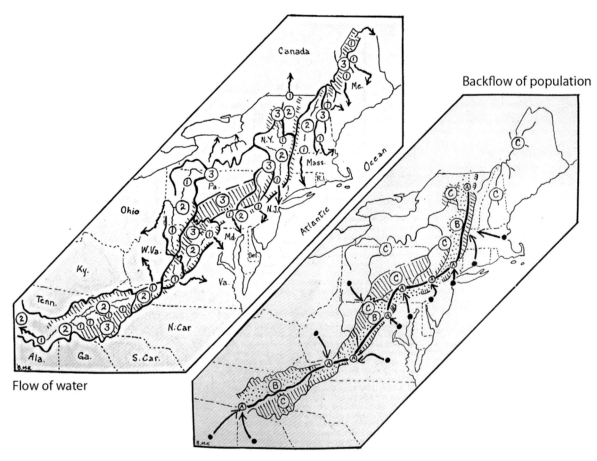

Figure 10.4 Benton MacKaye's diagrams of water and population 'flows'

Survey Graphic no. 22, 1933, pp. 252–253.

Figure 10.5 Norris Freeway, 1935
Tennessee Valley Authority/National Archives, Washington, DC.

Figure 10.6 Sketch of Norris Dam, probably drawn by Roland Wank (n.d.)
Tennessee Valley Authority/National Archives, Washington, DC.

people through the buildings and over the dams. Locks allowed goods to flow up and down the river. Even electricity was a flow. The aim was to direct and harness these flows, reversing the waste of natural and human resources wrought by centuries of neglect.

During his tenure at the TVA, MacKaye stayed a 'big picture thinker'. His experience and breadth of vision inspired the young planners, who met with him regularly after working hours, to talk about the ideas underlying their regional plan. His influence can be seen in the design of Norris Freeway, the regional parks developed around the reservoir, and in the image of Norris Town as a modern but rural community. His image of the highway as a levee to control urbanization, was welcomed by TVA planners. In the words of Earle Draper, the director of TVA planning,

no unsightly billboards, filling stations [...] or other eyesores are permitted along its margins. [... It] has been designed as a natural development [...] literally molded into the earth. [...] Instead of being just another ribbon of concrete, it flows along and around the slopes of the ground and through the valleys and hollows as naturally as a mountain stream finding its way to the river.[23]

In this sense, the highway at Norris was a naturalized artery, cleared of obstructive accretions in order to facilitate easier circulation of the economic lifeblood of the region. It was designed to follow the flow of the land, like a natural waterway, leaving the elevated landscapes to the hiker and the forest wildlife.[24]

CONTACT WITH NATURE

In their land use planning of conservation areas, the TVA focused equally on functions and flows. Reforestation and repair of eroded landscapes were critical factors in realizing the dams' function of flood control, so the TVA purchased

of flowing water into electricity; but in their design, the flow of people moving through, across and around the dams was equally considered. Visitor centres were designed to circulate and consolidated a continuous, and sizable, tract of land around each new reservoir. This was a costly and politically difficult process that often led to litigation. These areas were further articulated into conservation and recreation areas.

The first stage in developing these conservation areas was the introduction of Civilian Conservation Corps (CCC) camps, established after work had begun on the town of Norris. The CCC was a make-work program that brought young out-of-work men from cities such as New York to wilderness areas, employing them in building park structures and roads, reforestation projects, and erosion control. It was also a direct example of MacKaye's vision for the 'backflow' from the metropolis to rural areas, first outlined in his Appalachian Trail essay of fifteen years earlier. The CCC units assigned to the Norris reservoir were put to work creating parks along the shores of the future lake.

Their efforts resulted in parks that were more oriented towards recreation than the wilderness experience, although the degree to which the wilderness ideal or the recreational model for parks would prevail was a matter of disagreement in the TVA. Some argued that there should be no recreational structures of any kind in the parks around the dams. Others saw the recreational potential of these areas as a key selling point for the TVA to persuade the public of the value of land appropriations.

A HARMONIOUS BALANCE

The ultimate goal of flow control was to allow people to dwell in harmony with nature. To do this, houses, farms, manufactures, and parks all had to be linked in a system of exchanges so people could contribute to national renewal. This leads us to the culminating image of MacKaye's Appalachian Trail essay as it finds its development in the TVA: his view of rural culture as the well-spring of authentic life.

Figure 10.7 Conceptual drawing of Norris Town, 1933
Tennessee Valley Authority/National Archives, Washington, DC.

Figure 10.8 Aerial view of Norris Dam, ca. 1936
Tennessee Valley Authority/National Archives, Washington, DC.

In *The New Exploration*, he charts out three stages in the historical development of American culture: first, the outward exploration of the frontier – personified in explorers, frontiersman, and settlers. This is followed by a second stage, the inward exploration of spirit, found in the writings of Ralph Waldo Emerson, Walt Whitman and Henry David

Thoreau – all authors MacKaye greatly admired, and ones whose insights into the human condition owed a large debt to their contact with wilderness. In this section, MacKaye focused on the inner exploration triggered when people encounter nature.

At the dawn of the twentieth century, wrote MacKaye, the United States has found itself in crisis, as the inexorable spread of civilization submerges the territory of Thoreau as surely as it overwhelms that of the frontiersmen. The resulting 'wilderness of civilization' demands a new exploration into how it might be tamed and controlled to serve human purposes. 'The contest is between metropolitan America and indigenous America' and the solution is the regional city – which combines urbanity, rusticity and the primeval world of nature.

We turn here to the town of Norris, in its setting of flows and conserved parklands. Norris was the first TVA project, and its town was built to house the engineers, workers, and administrators building the dam. It was an early example of greenbelt town planning, weaving footpaths and roads through rolling hills and forest. A school, shops and a cooperative grocery were near the town commons, the experimental farms and small manufactures on its periphery. It was designed to balance commerce and industry with rural life and wilderness, and it was envisioned as a model for new towns across the region.

In Norris, the farming and craft activities of rural life were integrated with the goods and commerce of a small town, modernized by electricity, and integrated into a cycle of flows that extended beyond regional boundaries. MacKaye provided the theoretical framework for this focus on flows, as it was developed in Norris Town (separating pedestrians from cars) and, at a larger scale, by how the town fit into a larger system of settlements, dams and parklands linked by park-lined highways into a seamless whole.

In closing, let us return to the view from above. In his book *Utopics: Spatial Play,* Louis Marin describes the operation of such a view:

[such a] viewpoint is fixed at a totalizing point of view. One can see all. But the eye placed at this point occupies a place that is an 'other' point of view: it is in fact impossible to occupy this space. It is a point of space where no man can see: a no-place not outside space but nowhere, utopic.[25]

And it was indeed a utopian vision – or, we might say, a planner's vision – in which nature, technology, and people had to be accommodated. Nature set the parameters for the scope of the project; from the beginning the TVA realized that the unit of planning was the Tennessee watershed. Technology, in the forms of dams and highways, would be designed to fit seamlessly into the surrounding landscape. Working as valves and conduits, dams transformed natural flows into power that helped people farm more effectively, with fertilizers, irrigation pumps, or incubators. And finally, people had a central place in this renewed natural environment. The new arcadia would reunite people and technology in nature, and with nature – the improvement of the biological world being the ultimate goal. Again, we return to MacKaye:

We discover a thing by losing it. Since 1908 we have discovered this new resource of environment – something as 'new' as air would be if it were gradually turned off. Air is to the body what environment is to the mind – a fundamental medium and resource. We are losing this medium and resource – on roads, in towns, in mountain spaces.[26]

The ultimate expression of the arcadian ideal restructured for a technological age exited only in a utopian interval – that is, the first six years of the TVA – in the comprehensive plan for Norris. As the TVA moved its efforts downstream, constructing a dozen other dams, this vision collapsed under the weight of political realities and ideological differences in its leadership, and ultimately, in 1939, ran into the exigencies of war. Yet MacKaye's advocacy of the importance of nature, and his obsession with flows, can be seen today in the parks and rolling routes of the TVA system, a reconstructed nature knit into state and national parks with inspiring overlooks and a breath of fresh air.

ENDNOTES

1 See Daniel Schaffer, 'Benton MacKaye: the TVA years', *Planning Perspectives* 5 (1990).

2 David E. Nye, 'Redefining the American Sublime, from Open Road to Interstate' in this volume.

3 See Donna Haraway, 'Teddy Bear Patriarchy in the Garden of Eden', *Primate Visions: Gender, Race, and Nature in the World of Modern Science* (London: Routledge, 1989).

4 First articulated by Frederick Turner in his 'frontier hypothesis' of 1893, a seminal influence on the conservation movement championed by Theodore Roosevelt and others, and on the establishment of the National Park system. See Christine Macy and Sarah Bonnemaison, *Architecture and Nature: Creating the American Landscape* (London: Routledge, 2004).

5 Paul Sutter, 'A Retreat from Profit: Colonization, the Appalachian Trail, and the Social Roots of Benton MacKaye's Wilderness Advocacy', *Environmental History* 4, 4 (October 1999), p. 553.

6 Benton MacKaye, Speech to the Green Mountain Club, 'Why the Appalachian Trail?' 12 January 1929.

7 Benton MacKaye, *Project for an Appalachian Trail*, pp. 326–327.

8 Sutter, 'Retreat from Profit', p. 562.

9 With some successes. For example, the 170 km-long Skyline Drive in Shenandoah National Park (1931–39), connected to the 750 km-long Blue Ridge Parkway (1935–87). See Thomas Zeller, 'Staging the Driving Experience: Parkways in Germany and the United States' in this volume.

10 Benton MacKaye, *The New Exploration: A Philosophy of Regional Planning* (New York: Harcourt, Brace and Company, 1928). See also Benton MacKaye, 'The New Exploration: Charting the Industrial Wilderness', *Survey Graphic* 54 (1 May 1925), pp. 153–157, p. 192, p. 194.

11 Benton MacKaye, 'The Townless Highway', *The New Republic* (12 March 1930), p. 94. See also R.D. MacKenzie, Review of *The New Exploration* by Benton MacKaye, *The American Journal of Sociology* 35, 4 (January 1930), pp. 668–669.

12 MacKaye, *The New Exploration*.

13 Rebecca Conard, Review of *Regional Visionaries and Metropolitan Boosters: Decentralization, Regional Planning, and Parkways during the Interwar Years* by Matthew Dalbey, *Environmental History* 9, 1 (January 2004), p. 149.

14 Benton MacKaye and Lewis Mumford, 'Townless Highways for the Motorist', *Harper's Magazine* 163 (August 1931), pp. 347–365.

15 Daniel Schaffer, 'Benton MacKaye', p. 9. Other Schaffer articles on the fate of the regional planning ideas of MacKaye and his like-minded colleagues in the TVA include 'Ideal and Reality in 1930s Regional Planning: The Case of the Tennessee Valley Authority', *Planning Perspectives* 1 (1986), pp. 27–44; and 'Environment and TVA: Towards a Regional Plan for the Tennessee Valley, 1930s', *Tennessee Historical Quarterly* 4 (1984), pp. 333–354.

16 He first proposed that the construction access road to Norris Dam be built as a demonstration of a limited access 'freeway', allowing for rapid movement (35 mph), curved to follow the contour of the rolling landscape, and provided with a 250-foot right of way to prevent the erection of billboards or unsightly roadside shacks. [Schaffer, 'Benton MacKaye', p. 15.]

17 Earle S. Draper, 'The New TVA Town of Norris, Tennessee', *American City* 48 (December 1933), p. 67.

18 Tracy Augur, 'The Planning of the Town of Norris', p. 21.

19 In distinction to the 'parkways' discussed by Zeller and Nye in this volume, on which truck traffic was forbidden.

20 Benton MacKaye, 'The Challenge of Muscle Shoals', *The Nation* (19 April 1933), p. 446.

21 Schaffer, 'Benton MacKaye', p. 10.

22 Benton MacKaye, 'Tennessee – Seed of a National Plan', *Survey Graphic* 22 (May 1933), pp. 251–254.

23 Earle S. Draper, 'The TVA Freeway', *American City* 49, 2 (February 1934), pp. 47–48.

24 According to Thomas Zeller, MacKaye argued that highways in mountain regions should be designed to follow valleys, protecting the ridges as wilderness zones. See Zeller, 'Staging the Driving Experience' in this volume.

25 Louis Marin, *Utopics: Spatial Play* (Atlantic Highlands, NJ: Humanities Press; London: Macmillan, 1984), p. 207.

26 MacKaye, 'Challenge of Muscle Shoals', p. 445.

11   Aerocabs and Skycar Cities: Utopian Landscapes of Mobility

Even Smith Wergeland

The banker's aerocab had just docked at the belvedere landing-platform on the top of the mansion. Leaving the vehicle to his servants, he took the elevator downstairs.[1]

The above quotation is taken from *The Twentieth Century*, the first volume of a trilogy of illustrated novels created by the French science fiction maverick Albert Robida between 1883 and 1890.[2] Robida embraced science fiction both as a written genre and as a subject of illustration, a combination that enabled him to visualize his own tale of a future Paris. His peculiar visual universe takes form from pencil and ink drawings and a variety of graphical printing techniques, from hand-coloured lithographs to simple black and white compositions. The artwork makes up a fascinating portfolio of icons of the modern machine age, ranging from inventions that later became part of reality, for example a helicopter prototype, to ravings like aerocabs and flying caravans.

The anticipatory nature of Robida's work adheres to what Fredrick Jameson calls 'Utopian Futures'; conceptions of the future based on imaginary, fictive proposals.[3] This essay explores certain configurations of utopian futures within architectural and popular culture, in particular the desire to depict and design future landscapes by means of pre-emptive aesthetic proposals.[4] To limit this rather wide topic I specifically deal with representations of imaginary infrastructural schemes, i.e. landscapes of mobility, as I trace the history of the future from Robida's work in the late nineteenth century to the elaborate future scenario of MVRDV's *Skycar City* project from 2007.[5]

DESIGNING UTOPIAN FUTURES

Utopian visions usually feature re-modulations of city spaces and infrastructure. One could however ask to what extent they succeed in exercising any influence on our abilities to create worthwhile visual conceptions of the future. This question was explored at the 1939/40 New York World Fair through a vast number of explorative models of urban landscapes obviously influenced by science fiction discourses like Robida's. One of the biggest attractions at the Fair was the exhibition called *Futurama*, a massive assembly of models and billboards created by the prolific American industry designer Norman Bel Geddes, who was financially supported by General Motors for the occasion. Through this 'Autopia', as Robert Fishman calls it,[6] we are presented with a prototype for a United States to come. Bel Geddes' models project the viewer 20 years into a future, where the American landscape is overtaken by the splendours of motorization. *Futurama* showcases a world of vast expressways, sculpturally formed traffic machines, slender flyovers, and towering skyscrapers:

Figure 11.1 Albert Robida, Airship floating above the mountains, *ink graphite on pencil drawing, 1883*

Robida's trilogy *Le Vingtième Siècle* (Paris: Decaux, 1883).

a tight symbiosis of roads and high-rise buildings. It is impossible not to associate Bel Geddes' imagery with Le Corbusier's work in the 1920s and 30s. With urban scenarios like *Plan Voisin* Le Corbusier intended the automobile to save the city, as he writes in *The City of To-morrow and its planning,* and he made use of the skyscraper as a vertical reflection of the horizontal shape of urban roads, thus introducing the idea of 'streets in the air'.[7]

Both Bel Geddes and Le Corbusier were obsessed with the aesthetics of speed and designed their cities accordingly,

as landscapes where mobility reigns at every level. In both cases, however, the visions remain unfulfilled. Le Corbusier's fascination for many-layered transport systems could not disguise the difficulty of actually building streets in the air in an actual city. Although their infrastructural schemes exceeded most of their contemporaries in terms of visionary boldness, Bel Geddes and Le Corbusier's proposals were in fact intended as projections of the future based on available technology and attainable goals. Bel Geddes himself summed up the desired impact of *Futurama* with the following words: 'The plan it presented appealed to the practical engineer as much as to the idle day-dreamer. The motorways which it featured were not only desirable, but practical'.[8] Given that *Futurama* was one of the most popular exhibits at a world fair that was generally considered a great success, the legacy of Bel Geddes' utopian future should not be underestimated. [9]

What, then, about utopian futures that venture beyond the realm of reality? As we now move back to the future, to borrow one of the most chronologically confusing titles in film history, we shall see that the cityscapes of Albert Robida in many ways surpass Bel Geddes and Le Corbusier when it comes to sheer imaginary power.

INVENTING THE UTOPIAN LANDSCAPE OF MOBILITY

The second half of the nineteenth century saw the emergence of science fiction as a full-blown genre in literature, for example by the work of Jules Verne and H.G. Wells.[10] Science fiction quickly became a vehicle for the creative imagination of artists and popular magazine illustrators during this period. Out of this current came Robida, who in his trilogy catapults the reader approximately seventy years into the future. A driving force throughout the whole trilogy is everyday scenarios altered by puzzling technological devices. The *Aerial Rotating House,* for instance, is a compelling technological wonder,

Figure 11.3 Albert Robida, Leaving the Opera, *lithograph, 1883*
Robida's trilogy Le Vingtième Siècle (Paris: Decaux, 1883).

Figure 11.2 Albert Robida, Aerial Rotating House, *ink graphite on pencil drawing, 1883*
Robida's trilogy *Le Vingtième Siècle* (Paris: Decaux, 1883).

a traditional Parisian house turned into a mobile dwelling (Figure 11.2). This image is representative for Robida's ability to integrate wondrous scientific inventions into everyday life. It is mainly because of this mixture of the ordinary and the extra-ordinary that some commentators have dubbed Robida's output as both futuristic and traditional at the same time.[11] There is one particular aspect of his images that justifies this characteristic. His Paris cityscapes consist exclusively of historical architecture, while the machine park and other inventions are truly futuristic. In *Leaving the Opera*, Robida's high-flying solution to the logistics of bourgeois culture, the characters are dressed as citizens of the nineteenth century but the vehicles they are boarding do not belong in that context.

Another observation worth noting is the idea of movement above street level. Visions of flying vehicles and multilayered routes of transportation frequently appear in his drawings. Movement is sprawling at all levels of the city as transportation vehicles and technological auxiliaries aid the inhabitants. Robida draws up a visual structure governed by the dynamics of speed: rapid movement above the city and slower tempo at ground level, just as Le Corbusier envisioned the structure of the city of tomorrow. However, Robida's fictional world of mobility isn't simply about movement through transportation. It also reflects on social movements such as the women's liberation movement. In Robida's 1950s Paris women play crucial roles in society. Throughout the whole trilogy they are presented as political leaders, lawyers and officers in the army.

An interesting aspect of Robida's utopian future is that it is visualized through an already existing environment. Instead of rearranging the cityscape and introducing new architectural forms, he foresees a Paris of the future in an almost perfectly preserved state. It is Robida's machinery of transport and movement and the social mobility prospects that give the future a whole new dimension. The future lies, so to speak, in mobility itself. The citizen's ability to move about at dramatically increased speeds is what drives the action of the whole trilogy. This event from the opening chapter really sets the scene:

A train from the Brittany Tube had just docked, its passengers quickly filling up a dozen aircraft hovering over the station. This unloading of these travellers had also generated a swarm of smaller vehicles: fully-loaded aerocabs, veloces, pinnaces, flashes and freight tartans.[12]

With descriptions like this, aided with the striking force of the images, Robida paved way for the modern conception of utopian landscapes of mobility and, as we soon shall see, his work encapsulated elements that gained real momentum many decades later.

CIRCULATING THE FUTURE

There is perhaps no direct link from Robida to Le Corbusier and Bel Geddes. However, during the time-span that separates them, a vigorous exchange of ideas went on between the fields of magazine illustrations and architectural drawing. An interesting example is the case of the Italian architect Antonio Sant'Elia, who was an affiliated member of the Italian Futurist Movement. According to the American architectural historian Esther da Costa Meyer, Sant'Elia was deeply inspired by popular magazines when he came up with several of his grand urban schemes.[13] In fact, da Costa Meyer claims that Sant'Elia's projects owed more to magazine illustrations as sources of inspiration than to contemporary architectural projects, such as Eugène Hénard's *Cities of the Future* from 1910.[14] In her book on Sant'Elia, *The Work of Antonio Sant'Elia – Retreat into the Future*, Meyer conducts a thorough analysis of a range of Sant'Elia's sketches. She compares them to the work of the American illustrator Charles Lamb, whose drawings of cities with many-layered transportation schemes were published in the Italian magazine *L'Illustrazione Italiana* and several other European magazines in the early nineteen hundreds.

L'Illustrazione Italiana also introduced Sant'Elia to the work of the Italian architect Emilio Belloni, who came up with a scheme called *Corso d'Italia* [The Italian Road]: a 260 feet wide highway that penetrates Milan from the heart of the city, right next to the Duomo, and ends nineteen miles away. Belloni's work was illustrated by the young architect Mario Stroppa, whose exuberant charcoal drawings provided Sant'Elia with visually convincing answers to urgent needs. He appropriated the visionary boldness of Lamb and Belloni in projects like *Station for airplanes and trains* and *Milan, Year 2000*, a fully-fledged master plan for Milan. None of these projects were executed, a fate they shared with the majority of Sant'Elia's proposals. The important point in this context is that his work bridges the gap between architecture and popular culture, answering Meyer's assertion that 'Science fiction created a certain climate for euphoria about the future and popularized ideas that were rapidly becoming realities'.[15]

This euphoria about the future resurfaced in the post war period, when topics from science fiction and popular magazines became widely circulated, fuelled by the rise of youth culture and pop art. A typical outcome of this cross-modal cultural circulation is The Independent Group, an association of artists and architects who were inspired by new movements in art and science fiction. In 1956 the group put up an exhibition called 'This Is Tomorrow' at London's Whitechapel Art Gallery. The exhibition was a collaborative effort merging architecture and art with icons of science fiction and objects from popular culture. Among the contributors was the young architect couple Alison and Peter Smithson, whose immense fascination with the concept of mobility became a trademark of the infrastructural utopias they produced in the late 1950s. Their theoretical position was completely governed by this mode of thought:

> Mobility has become the characteristic of our period. Social and physical mobility, the feeling of a certain sort of freedom, is one of the things that keep our society together, and the symbol of this freedom is the individually owned motorcar. Mobility is the key both socially and organizationally to town planning, for mobility is not only concerned with roads, but with the whole concept of a mobile, fragmented, community.[16]

The Smithsons' insistence on mobility as the most influential phenomenon in society, as the above quotation so explicitly maintains, is a critical attack on the pre war modernist dogma. They are fiercely bent on revealing the inadequacies of Le Corbusier's urbanism, which is labelled too static and rigid to handle the complexities of a society in rapid transformation. Despite their critique of the previous generation of architects and planners, their reassessment of modern architecture in many ways connects with the urban landscape utopias produced in the past. It would be hard to imagine the Smithsons' radical aesthetics of mobility unrelated to preceding projects like *Broadacre City*, Frank Lloyd Wright's vision of decentralized modern living made possible by the qualities of the motorcar and a flexible infrastructural system. Interestingly enough, the various illustrations of the *Broadacre City* plan include both ordinary roads for motor traffic, displayed as two-lane motorways cutting through the city in orderly fashion, and some remarkably futuristic transport networks: semi-circular, mushroom-shaped flying vehicles hover above the landscape while curious-looking passenger cars or

motorcycles dart about at ground level. Robida's aerocabs certainly haunt the images of *Broadacre City*, inter-textually speaking. Wright's peculiar vehicles reaffirm the flying car as an indispensable icon of the future. The same goes for one of Wright's contemporaries, Buckminster Fuller, who launched the Dymaxion car in 1933, one year after Wright's publication of *Broadacre City*.

The Dymaxion car never actually flew and Wright's utopian vehicles remain as unrealized as the visionary city they were invented for. But Buckminster Fuller and Wright's ideas persisted. The significance of their output is summed up this way by the American scholars Peter Wollen and Joe Kerr: 'The car-based future championed by the Modernist avant-garde may not have come to pass exactly as they envisioned, but it happened all the same'.[17] Here, Wollen and Kerr allude to the fact that many European and American cities were overtaken by large motorway networks in the post war period. Meanwhile, as car culture gradually gained control over the urban landscapes of the real world, new conceptions of the utopian landscape of mobility were developed. In America, images from Bel Geddes' *Futurama* were appropriated by the growing B-film industry, as seen in the American painter Reynold Brown's poster for the movie *Attack of the 50 ft Woman* from 1958. Brown's artwork and similar efforts continued to negotiate the future as a landscape of mobility.

These examples demonstrate how the particular idea of the future as mobility constantly reoccurs due to an extensive circulation of visual images. This circulation produces an urban landscape iconography in which utopian images from different contexts merge because of their visual inter-textuality; their *interbildlichkeit*, to borrow a term from Margaret Rose.[18] Although the examples are separated by historical time and cultural context, they form a continuous discourse held together by the practise of visualizing mobility. Robida's conception of the future never seems to wear off as new interpretations of his ground formula frequently appear.[19]

THE TWENTY-FIRST-CENTURY UTOPIAN LANDSCAPE OF MOBILITY

Roads? Where we're going, we don't need roads.[20]

In 2007, nearly 120 years after Robida's seminal work, the Dutch architectural practise MVRDV published the results of an experimental studio project called *Skycar City*. In their introduction the architects explain the potential outcome: 'This [project] will lead to a city of "streets" at any level, or perhaps empty of streets as we know them'.[21] Le Corbusier's concept of 'streets in the air' resonates loudly in this formulation, as do the infrastructural terminology of the radical agents of post-war architectural discourse, such as the Smithsons and Archigram. The book contains an eclectic list of references, spanning from the technological experiments of Leonardo Da Vinci, via Le Corbusier's expressways to the popular American sci-fi movie *Back to the Future* from the 1980s. By activating these influences the architects feed upon the history of the future in a highly conscious way: 'Skycar City is positioned at the confluence of all these issues, speculating on urban transportation infrastructure as a medium through which to imagine the potentials of the new city, one where movement is both efficient and free'.[22]

With this statement as a starting point, the hyper-volumetric world of *Skycar City* tries to reinvent the various historical impulses in a contemporary mode. The outcome of this brief is tested in two different scenarios: a purely hypothetical one and a version of the project implemented in the city of Tianjin in China. The question of how and to what degree transportation can shape architecture, with one constant variable – the car – lies at the core of both versions. This clearly relates to the tricky question of reality and fiction. *Skycar City* is embedded within the imaginative force of science fiction but is consolidated within the profession of architecture and urbanism. It is perhaps best described as informed science fiction, a knowledge-based speculation in touch with real life.

The architects themselves openly agree that *Skycar City* is funded on wishful thinking to a significant extent since they knowingly have deployed imaginative prophecies as generators for the project. But still, they do try to adapt it to an existing setting and the whole presentation answers to conventions one would expect from any serious architectural competition entry. *Skycar City* has a program arrangement and the architects use words like 'probability' and 'application' to define the design parameters. However hypothetical the basic ideas may be, one gets the feeling that *Skycar City* is actually achievable because of the convincing systematization

Figure 11.4 MVRDV, Skycar City docking system, *digital rendering, 2007* MVRDV's Skycar City (Barcelona, 2007). Reproduced with permission of MVRDV.

of different ideas. Also, it manages to incorporate relevant thoughts about landscape and infrastructure that have often been excluded from the discourse of architecture.

The last point is proven, as it were, by the considerable similarity between *Skycar City* and Robida's future trilogy. MVRDV does have the courtesy to acknowledge Robida's contribution in making the skycar an icon of the future by crediting him as the originator of the concept.[23] There are quite a number of direct connections between *Skycar City* and Robida's vision for Paris. The docking system of *Skycar City* equals Robida's landing platforms (Figure 11.4). A significant amount of traffic takes place in the air and the speed structures are almost identical – fast acceleration above the city, slow tempo at ground level. The biggest difference

Speed Series 360°

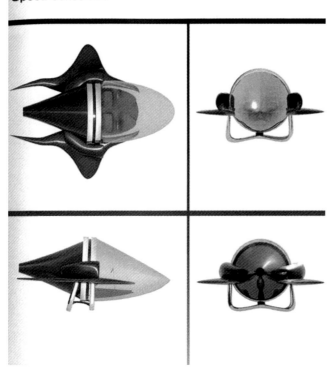

Figure 11.5 MVRDV, The Speed Series, *digital rendering, 2007*
Reproduced with permission of MVRDV.

between them is that *Skycar City* consists of brand new architecture and that contemporary trends in city planning are implemented, for instance environmental-friendly solutions. According to the architects' prophetic calculations, skycars will replace ground-based vehicles within 2030 and by 2075 oil and gas will be entirely replaced by wind, solar and tidal energy systems.[24] MVRDV are prepared for any challenge and they have the necessary equipment ready at hand: the Skycar Collection, a car fleet of vehicles designed to meet the demands the future.

The inter-textual journey from Robida's aerocabs to MVRDV's *Skycar City* indicates at least two valid interpretations. One the one hand the examples in question project the future as a prosperous landscape of mobility and fanciful technologies. This process of inter-connectivity and continuous recirculation links these landscapes of mobility and constitutes a diverse iconography of the future. Despite their imaginative nature, the majority of these visions unfold in realistic environments. They are proposed as operative possibilities and intentions for the future and cannot be discarded as mere utopias. Rather, their utopian character aspires to Jean Baudrillard's idea of fiction as a phenomenon that 'anticipated imagination by giving it the form of reality'.[25] In this regard they serve as powerful sources of inspiration and cues to a vibrant visual practise with great relevance for how we understand reality. Through this practise the concept of mobility is ascribed new meanings through an ever-evolving discourse of paraphrasing and remodelling. Mobility is reaffirmed as an important occurrence in contemporary culture as it manifests itself in a range of vivid futuristic landscapes.

One the other hand, if one accepts the interpretation of these images as being part of a coherent iconography, one might argue that they constitute a discourse of stand-still and creative drought. 'The future ain't what it used to be', the legendary American baseball coach Yogi Berra once declared. But maybe the future is, in fact, exactly what it used to be. In spite of the diversity in expression, the landscape iconography of the future is founded on a set of common denominators; recurring elements like flying taxis, multi-layered traffic, and rapid transportation. The future is always presented as fast, free flowing and efficient. Moreover, depictions of the future have always been concerned with vehicles situated in the higher spheres, from Robida's aerocabs to the zeppelins of *King's View of New York* by Moses King.[26] Therefore, one could easily say that most of the ingredients of *Skycar City* are evident and conventional. Despite the fact that it entails many new elements compared to Robida it is basically the same old formula in repeat. Seen this way, the idea of the future as a utopian landscape of mobility is first and foremost initiated through a static reproduction of conventionalized concepts.

This aspect is addressed by Fredric Jameson in his comment on the concept of Utopia in science fiction. In *Archaeologies of the Future* Jameson argues that even contemporary visions of the future have become historical and dated: 'Utopian future has turned out to be merely the future of one moment of what is now our own past'.[27] Baudrillard also admits that 'We are condemned to the imaginary and to nostalgia for the future'.[28] Utopian landscapes of mobility constitute a paradox. They fetishize mobility and unbridled movement but remain at the same time quite stationary cultural conceptions. Their visually seductive qualities continue to fascinate us and trigger new infrastructural visions, but the future remains the same. Their actual influence on our conception of urban landscapes is, it seems, more stifling than stimulating.

ENDNOTES

1   Albert Robida, *The Twentieth Century* (Middletown: Wesleyan University Press, 2004), p. 10.

2   The trilogy was originally published as *Le Vingtième Siècle, La Guerre au vingtième siècle* and *Le Vingtième Siècle. Le vie èlectrique* (Paris: Georges Decaux, 1883, 1887, 1890).

3   See Fredric Jameson, *Archaeologies of the Future – The Desire Called Utopia and Other Science Fictions* (London/New York: Verso Books, 2005).

4   I borrow the term 'pre-emptive' from the title of MVRDVs Skycar City publication because it fits well with all the selected examples, which relate to one another by their energetic projection of the future. See Winy Maas & Grace La (eds), *Skycar City – a Pre-emptive History* (Barcelona: Actar, 2007).

5   In the context of this chapter, the term 'landscapes of mobility' refers predominantly to landscape representations characterized by movement, speed and circulation.

6   Robert Fishman, *Urban Utopias in The Twentieth Century* (New York: Basic Books, 1977), p. xiii.

7   Le Corbusier, *The City of To-morrow and it's Planning* (London: The Architectural Press, 1971), p. 55.

8   Norman Bel Geddes, *Magic Motorways* (New York: Random House, 1940), p. 6.

9   For a brief summary on the impact of the New York World Fair, see Donna Goodman, *A History of the Future* (New York: The Monacelli Press, 2008), pp. 118–119.

10 Jameson, *Archaeologies of the Future*, p. 284.

11 Philippe Willems, 'Introduction', in Albert Robida, *The Twentieth Century* (Middletown: Wesleyan University Press, 2004), p. 17.

12 Albert Robida, *The Twentieth Century*, p. 3.

13 Esther da Costa Meyer, *The Work of Antonio Sant'Elia – Retreat into the Future* (New Haven: Yale University Press, 1995), pp. 128–132.

14 Ibid, p. 128.

15 Ibid, p 137.

16 Alison and Peter Smithson, 'Mobility', in *Architectural Design*, vol. 126 (October 1958), p. 386.

17 Peter Wollen and Joe Kerr (eds), *Autopia – Cars and Culture* (London: Reaktion Books, 2002), p. 235.

18 See Margaret A. Rose, *Parodie, intertextualität, interbildlichkeit* (Bielefeld: Aisthesis Verlag, 2006).

19 A curious example is the Japanese animation movie *The Invention of Destruction in the Imaginary Machines* from 2002, which paraphrase's Robida's fictional world of flying taxis and vehicles shaped like fish hovering above a historic-looking cityscape.

20 This is uttered by the character Doc Emmett Brown (played by Christopher Lloyd) in the final scene of *Back to the Future*, seconds before the protagonists take a flying DeLorean into the future. See Robert Zemeckis (dir.), *Back to the Future* (Hollywood: Universal Pictures, 1985).

21 Winy Maas and Grace La, *Skycar City – a Pre-emptive History*, p. 4.

22 Ibid, p. 14.

23 Ibid, p. 2.

24 Ella Peinovic, 'Designing SkyCar City: A Post-Studio History', http://thewhereblog.blogspot.com/2008/03/designing-skycar-city-studio.html, p. 5.

25 Jean Baudrillard, *America* (London: Verso Books, 1988), p. 95.

26 See Moses King, *King's Views of New York, 1896–1905, & Brooklyn, 1905* (New York: Arno Press, 1977).

27 Jameson, *Archaeologies of the Future*, p. 286.

28 Baudrillard, *America*, p. 95.

Section III   Landscapes of Mobility

Mobility has been invested with a multitude of cultural and aesthetic significances in the twentieth century and continues to be a pressing issue in both academic and architectural discourse. This section investigates the particular mobility at stake in roads and road installations, looking specifically at the road as an aesthetic orchestration of movement and rest.

12   Towards a Politics of Mobility

Tim Cresswell

The last few years have seen the announcement of a 'new mobilities paradigm',[1] the launch of the journal *Mobilities* and a number of key texts and edited collections devoted to mobility.[2] Work inspired by the 'new mobilities paradigm' has informed a diverse array of work on particular forms and spaces of mobility ranging from driving and roads[3] to flying and airports.[4] This is not the place to review the work on mobility.[5] Rather, the overall aim of the chapter is to move forward with some of the insights of the mobility turn, or new mobilities paradigm, and further develop some of the ideas that have been associated with it.[6] In particular, this chapter develops the approach I utilized in *On the Move: Mobility in the Modern Western World*. In that book I outlined the role of mobility in a number of case studies ranging in scale from the micro-movements of the body to the politics of global travel. But, for the most part, mobility remained a singular thing. There was no detailed accounting of various aspects of mobility that have the capacity to make it powerfully political. This chapter, then, is an attempt at outlining some key ideas for a meso-theoretical approach to the politics of mobility. Strategically, it uses ideas from other theorists and a variety of real world examples. It does not subscribe to a singular theoretical model but seeks to contribute to the development of a geographical theoretical approach to mobility. It is part of an ongoing process of meso-theoretical construction.

The chapter seeks to meet these aims in two principal ways. First, by breaking mobility down in to six of its constituent parts (motive force, velocity, rhythm, route, experience and friction) in order to fine-tune our accounts of the politics of mobility, Second, by developing the notion of 'constellations of mobility' as historically and geographically specific formations of movements, narratives about mobility and mobile practices; which reveal the importance of an historical perspective which mitigates against an overwhelming sense of newness in mobilities research. First, however, consider the notion of a 'new mobilities paradigm'.

THE NEW MOBILITIES PARADIGM?

Bruno Latour has suggested that there are only three problems with the term Actor Network Theory – and they are the words 'actor', 'network' and 'theory'.[7] A similar point could be made of 'new mobilities paradigm'. First of all the word 'paradigm' suggests the Kuhnian notion of normal science being transformed by sudden revolutions where what went previously is unceremoniously tipped into the junkheap of academic history.[8] We have to be careful about such implications. Any study of mobility runs the risk of suggesting that the (allegedly) immobile – notions such as boundaries and borders, place, territory and landscape – is of the past and no longer relevant to the dynamic world of the twenty-first century. This would be wrong and, to be fair, does not seem to be the point of advocates of the new mobilities paradigm where 'moorings' are often as important as 'mobilities'. The second problem concerns the different ways that 'new mobilities' can be read. If the emphasis is on the word 'new' then this suggests an old mobilities paradigm. If the emphasis is on the word 'mobilities' then this suggests that old paradigms were about the immobile or sedentary. The second of these options seems untenable because movements of one kind or another have been at the heart of all kinds of social science (and particularly geography) since their inception. In sociology notions of movement and mobility were central to the concerns of thinkers such as Georg Simmel and the Chicago School sociologists for instance.[9] If we think of geography there have been any number of sub-disciplinary concerns with things and people on the move ranging from Saurian concerns with origins and dispersals[10] through spatial science's fixations of gravity models and spatial interaction theory[11] and notions of 'plastic space'[12] to feminist approaches to daily mobility patterns.[13] Transport geography, migration theory, time geographies, geographies of tourism – the list is endless. The same could be said of anthropology. So the question that arises is, what is 'new' about the new mobilities paradigm?

Despite all the caveats above there clearly is something 'new' about the ways mobilities are being approached currently that distinguishes them from earlier accounts of movement, migration and transport (to name but three of the modes of mobility that have long been considered). If nothing else, the 'mobilities' approach brings together a diverse array of forms of movement across scales ranging from the body (or, indeed parts of the body) to the globe. These substantive areas of research would have been formally held apart by disciplinary and sub-disciplinary boundaries that mitigated against a more holistic understanding of mobilities. In addition, the approaches listed above were rarely actually *about* mobility but rather took human movement as a given – an empty space that needed to be expunged or limited. In migration theory movement occurred because one place pushed people out and another place pulled people in. So despite being about movement, it was really about places. Similarly transport

studies have too often thought of time in transit as 'dead time' in which nothing happens – a problem that can be solved technically. Mobility studies have begun to take the actual fact of movement seriously.

I have argued that mobility exists in the same relation to movement as place does to location[14] and that mobility involves a fragile entanglement of physical movement, representations and practices. Furthermore, these entanglements have broadly traceable histories and geographies. At any one time, then, there are pervading *constellations of mobility* – particular patterns of movement, representations of movement and ways of practicing movement that make sense together. Constellations from the past can break through into the present in surprising ways.[15] In addition they entail particular 'politics of mobility'. In general, though, I have not considered how mobility is made of interconnected elements such as speed and rhythm which merit separate consideration. Before moving on to six aspects of the politics of mobility it is necessary to define mobility as the entanglement of movement, representation and practice.

MOVEMENT, REPRESENTATION, PRACTICE

Consider, then, these three aspects of mobility: the fact of physical movement – getting from one place to another; the representations of movement that give it shared meaning; and finally the experienced and embodied practice of movement. In practice these elements of mobility are unlikely to be easy to untangle. They are bound up with one another. The disentangling that follows is entirely analytical and its purpose is to aid theory construction. Different forms of mobility research are likely to explore facets of any one of these. Transport researchers, for instance, have developed ways of telling us about the fact of movement, how often it happens, at what speeds and where. Recently they have also informed us about who moves and how identity might make a difference.[16] They have not been so good at telling us about the representations and meanings of mobility either at the individual level or at a societal level. Neither have they told us how mobility is actually embodied and practiced. Real bodies moving have never been at the top of the agenda in transport studies. Understanding mobility holistically means paying attention to all three of these aspects.

Physical movement is, if you like, the raw material for the production of mobility. People move, things move, ideas move. The movement can, given the right equipment, be measured and mapped. These measurements can be passed through equations and laws can be derived from them. This positivist analysis of movement occurs in all manner of domains. The physical movement of the human body has been extracted from real bodies and used to develop model mobilities for, amongst other things, sports therapy, animation and factory motion studies.[17] In cities transport planners are endlessly creating models of mechanically aided physical movement in order to make transport more efficient, or less environmentally harmful.[18] In airports and railway stations modellers have used critical path analysis to measure the time taken to get between two points and then reduce it.[19] So understanding physical movement is one aspect of mobility. But this says next to nothing about what these mobilities are made to mean or how they are practiced.

Just as there has been a multitude of efforts to measure and model mobility so there has been a plethora of representations of mobility. Mobility has been figured as adventure, as tedium, as education, as freedom, as modern, as threatening. Think of the contemporary links made between immigrant mobilities and notions of threat reflected in metaphors of flooding and swamping used by journalists and politicians.[20] Or alternatively the idea of the right to mobility as fundamental to modern western citizenship which is expressed in legal and governmental documents.[21] Consider all the meanings wrapped up in car advertisements or mobile phones. To take just one kind of mobile practice,

the simple act of walking has been invested with a profound array of meanings from conformity to rebellion in literature, film, philosophy and the arts.[22] Geographers, social theorists and others have been complicit in the weaving of narratives around mobility. We have alternately coded mobility as dysfunctional, as inauthentic and rootless and, more recently as liberating, anti-foundational and transgressive in our own forms of representation.[23]

Finally there is practice. By this I mean both the everyday sense of particular practices such as walking or driving and also the more theoretical sense of the social as it is embodied and habitualized.[24] Human mobility is practiced mobility that is enacted and experienced through the body. Sometimes we are tired and moving is painful, sometimes we move with hope and a spring in our step. As we approach immigration at the airport the way our mobility feels depends on who we are and what we can expect when we reach the front of the line. Driving a car is liberating, or nerve wracking, or, increasingly, guilt-ridden. Whether we have chosen to be mobile or have been forced into it affects our experience of it. Sometimes our mobile practices conform to the representations that surround them. We do, indeed, experience mobility as freedom as the airplane takes off and the undercarriage retracts. At other times there is a dissonance between representation and practice. As we sit in a traffic jam maybe. Mobility as practiced brings together the internal world of will and habit[25] and the external world of expectation and compulsion. In the end, it is at the level of the body that human mobility is produced, reproduced and, occasionally, transformed.

Getting from A to B can be very different depending on how the body moves. Any consideration of mobility has to include the kinds of things people do when they move in various ways. Walking, dancing, driving, flying, running, sailing. All of these are mobile practices. Practices such as these have played important roles in the construction of social and cultural theory, philosophy and fiction. Take walking for instance. We can think of the way Michel de

Certeau uses walking to examine the spatial grammar of the city that provides a pre-constructed stage for the cunning tactics of the walk:

> The long poem of walking manipulates spatial organizations, no matter how panoptic they may be: it is neither foreign to them (it can take place only within them) nor in conformity with them (it does not receive its identity from them). It creates shadows and ambiguities within them.[26]

This story about walking replicates a number of literatures in which the walker is held forth as an exemplar of rebellion, freedom and agency in the city – the pedestrian hero[27] or the flaneur.[28] Practices are not just ways of getting from A to B; they are, at least partially, discursively constituted. The possibility of walking is wrapped up in narratives of worthiness, morality and aesthetics that constantly contrast it with more mechanized forms of movement which are represented as less authentic, less worthy, less ethical.[29] And it matters where walking happens – the walk in nineteenth-century Paris is very different from the walk in rural Mali or the walk in the contemporary British countryside.

In addition to being a traceable and mappable physical movement which is encoded through representation, walking is also an embodied practice that we experience in ways that are not wholly accounted for by either their objective dimensions or their social and culture dimensions. Here the approaches of both phenomenological inquiry and forms of non-representational theory give insight into the walking experience.[30] Similar sets of observations could be made about all forms of mobility – they have a physical reality, they are encoded culturally and socially and they are experienced through practice. Importantly these forms of mobility (walking, driving etc.) and these aspects of mobilities (movement, representation and practice) are political – they are implicated in the production of power and relations of domination.

SIX ELEMENTS OF A POLITICS OF MOBILITY

By politics I mean social relations that involve the production and distribution of power. By a politics of mobility I mean the ways in which mobilities are both productive of such social relations and produced by them. Social relations are of course complicated and diverse. They include relations between classes, genders, ethnicities, nationalities and religious groups as well as a host of other forms of group identity. Mobility, as with other geographical phenomena, lies at the heart of all of these. The illustrations I use in what follows are designed to illuminate a variety of politics rather than privileging one over another. My point in this chapter is the development of a geographical understanding of mobility that can in turn inform theorizations of gender, ethnicity or any other form of social relation.

Mobility is a resource that is differentially accessed. One person's speed is another person's slowness. Some move in such a way that others get fixed in place. Examples of this abound. Consider the school run that allows women (for the most part) to enact an efficient form of mobility so often denied them. At the same time it impacts on the ability of children to walk to school and makes the streets less safe for pedestrians. There is little that is straightforward about such an entanglement of gender, age and mobility. Consider the opening up of borders in the European Union to enable the enactment of the EU mantra of free mobility. This in turn depends on the closing down of mobilities at the borders (often airports) of the new Europe.[31] Speeds, slownesses and immobilities are all related in ways that are thoroughly infused with power and its distribution.

This politics of mobility is enriched if we think about mobility in terms of material movement, representation and practice. There is clearly a politics to material movement. Who moves furthest? Who moves fastest? Who moves most often? These are all important components of the politics of mobility that can be answered in part by the traditional approaches of transport studies. But this is only the beginning. There is also a politics of representation. How is mobility discursively constituted? What narratives have been constructed about mobility? How are mobilities represented? Some of the foundational narratives of modernity have been constructed around the brute fact of moving. Mobility as liberty, mobility as progress. Everyday language reveals some of the meanings that accompany the idea of movement. We are always trying to get somewhere. No one wants to be stuck or bogged down. These stories appear everywhere from car advertisements to political economic theory. Consider the act of walking once again. The disability theorist Michael Oliver has suggested that there is an ideology of walking that gives the fact of walking a set of meanings associated with being human and being masculine. Not being able to walk thus falls short of being fully human. Popular culture tells us that 'walking tall' is a sure sign of manhood, medical professionals dedicate themselves to the quest to make those who can't walk, walk again. All manner of technologies are developed to allow people to walk. The effect of such an ambulatory culture, he tells us, can be quite devastating on those who are being 'treated'. As Oliver puts it 'Not-walking or rejecting nearly walking as a personal choice threatens the power of professionals, it exposes the ideology of normality and it challenges the whole rehabilitation exercise'.[32] Here mobility and, particularly the represented meanings associated with particular practices, is highly political.

Finally, and perhaps most importantly of all, there is a politics of mobile practice. How is mobility embodied? How comfortable is it? Is it forced or free? A man and a woman, or a businessman and a domestic servant, or a tourist and a refugee may experience a line of a map linking A and B completely differently. The fact of movement, the represented meanings attached to it and the experienced practice are all connected. The representation of movement can certainly impact on the experience of its practice. Think about Mexican immigrants in the United States for instance. Compare that to a member of a multi-national corporation jetting between world cities. Consider the image of a train with Pullman carriages steaming through the landscape of late nineteenth-century America. Here is a description from a journalist in the *Chicago News*:

The world respects the rich man who turned to be a globe-trotter and uses first class cabins and Pullman cars, but has inclination to look over his shoulder at the hobo who, to satisfy this so strong impulse, is compelled to use box-cars, slip the board under the Pullman or in other ways whistle on the safety of his life and integrity of his bones.[33]

Here we have exactly the same act of moving from A to B but completely different practices of mobility and sets of represented meanings associated with them. The globetrotter sits in plush velvet seats and chooses from extensive wine lists while the hobo travels close to death on a wooden plank precariously balanced on the same carriage's axels. The mobile subject 'globe-trotter' signifies a different world from the mobile subject 'hobo'. The narratives and discourses surrounding them both make their mobilities possible and impact upon these very different practices. Indeed, just fifty years earlier the subject identities of 'globe-trotter' and 'hobo' did not exist just as the Pullman carriage or the transcontinental railroad did not exist. These mobile spaces, subjects and practices were all entangled in that particular moment.

To recap, I want to develop an approach to human mobility that considers the fact of movement, the represented meanings attached to movement and the experienced practice of movement. Taking all these facets seriously, I argue, will help us delineate the politics of mobility. And this is important, as there seems little doubt that mobility is one of the major resources of twenty-first-century life and that it is the differential distribution of this resource that produces some of the starkest differences today.

But this argument is still more suggestive than specific. There remains the task of breaking mobility down into different aspects of moving that each have a role to play in the constitution of mobile hierarchies and the politics of mobility. In the process of breaking mobility down in this way we get some analytical purchase on how mobility becomes political. Below I outline six aspects of mobility that each has a politics that it is necessary to consider.

First of all *why does a person or thing move?* An object has to have a force applied to it before it can move. With humans this force is complicated by the fact that it can be internal as well as external. A major distinction in such motive force is thus between being compelled to move or choosing to move. This is the distinction at the heart of Bauman's discussion of the tourist and the vagabond.

> Those 'high up' are satisfied that they travel through life by their heart's desire and pick and choose their destinations according to the joys they offer. Those 'low down' happen time and again to be thrown out from the site they would rather stay in [...] If they do not move, it is often the site that is pulled away from under their feet, so it feels like being on the move anyway.[34]

Of course the difference between choosing and not choosing is never straightforward and there are clearly degrees of necessity. Even the members of the kinetic elite who appear to move so easily through the world of flows must feel obligated to sign in to airport hotels and book first class flights to destinations 12 time zones away. Nevertheless, this basic difference in mobilities is central to any hierarchy and thus any politics of mobility. To choose to move or, conversely, stay still, is central to various conceptions of human rights within the nation-state[35] and within 'universal' regimes.[36]

Second – *how fast does a person or thing move?* Velocity is a valuable resource and the subject of considerable cultural investment.[37] To Paul Virilio speed, connected to the development of military technology in particular, is the prime engine for historical development. In *Speed and Politics* and elsewhere he paints a picture of ever-increasing velocity overwhelming humanity. Even such apparently fixed things as territory, he argues, are produced through variable speeds rather than though law and fixity. He proposes a 'science of speed', or *dromology*, to help us understand our present predicament. The faster we get, Virilio argues, the more our freedoms are threatened:

> *The blindness of the speed of means of communicating destruction is not a liberation from geographical servitude, but the extermination of space as the field of freedom of political action. We only need refer to the necessary controls and constraints on the railway, airway or highway infrastructures to see the fatal impulse: the more speed increases, the faster freedom decreases.*[38]

At its extreme, speed becomes immediacy – the speed of light that Virilio claims is at the heart of globalization. This is the speed with which information can travel around the globe having profound impacts on relatively solid, relatively permanent, places.[39]

But speed of a more human kind is at the centre of hierarchies of mobility. Being able to get somewhere quickly is increasingly associated with exclusivity. Even in air travel – where, since the demise of Concorde, all classes of passenger travel at the same speed, those 'high up', as Bauman would put it, are able to pass smoothly through the airport to the car that has been parked in a special lot close to the terminal. In airports such as Amsterdam's Schiphol, frequent business travellers are able to sign up to the Privium scheme where they volunteer to have their iris scanned to allow biometric processing in the fast lane of immigration. This frees up immigration officials to monitor the slow lane of foreign arrivals who are not frequent business travellers. Speed and slowness are often logically and operationally related in this way. And it is not always high velocities that are the valued ones. Consider the slow food and slow culture movements. How bourgeois can you get? Who has the time and space to be slow by choice? As John Tomlinson has put it in relation to the Italian slow city movement, CittáSlow:

[...] CittáSlow, in promoting the development of small towns (of 50,000 inhabitants or less) represents the interests of a particular spatial-cultural constituency and related localized form of capital. In a sense them, and without being unduly cynical, [CittáSlow] could be seen as defending enclaves of interest, rather than offering plausible models for more general social transformation.[40]

For some, slowness is impossible. Consider the workers in Charlie Chaplin's *Modern Times*. In its famous opening scenes we see a line of workers at a conveyer belt tightening nuts on some unspecified element of a mass production line. The factory boss is seen reading the paper and enjoying a leisurely breakfast. This is interrupted only when he makes occasional demands for 'more speed' on the production line below. Here the principles of Taylorism are used by Chaplin to satirize the production of speed among workers through time and motion study. Here speed is definitely not a luxury. Rather it is an imposition experienced by those 'low down'.

Third – *in what rhythm does a person or thing move?* Rhythm is an important component of mobility at many different scales.[41] Rhythms are composed of repeated moments of movement and rest, or, alternatively, simply repeated movements with a particular measure. Lefebvre's outline of rhythm analysis as a method of interpreting the social world is richly suggestive. It brings to mind the more phenomenological conceptions of 'Place-Ballet' developed by David Seamon and recently re-incorporated into a geography of rhythms by Tom Mels.[42] But unlike Seamon, Lefebvre delineates how rhythms, such as those visible on any such city square, are simultaneously organic, lived and endogenous and exterior, imposed and mechanical. Frequently the exterior rhythm of rationalized time and space comes into contradiction with lived and embodied rhythm: 'Rhythm appears as regulated time, governed by rational laws, but

in contact with what is least rational in human being, the lived, the carnal, the body'.[43] Rhythm, to Lefebvre, is part of the production of everyday life, thus: 'rhythm seems natural, spontaneous, with no law other than its own unfurling. Yet rhythm, always particular (music, poetry, dance, gymnastics, work, etc.) always implies a measure. Everywhere there is rhythm, there is measure, which is to say law, calculated and expected obligation, a project'.[44] Rhythm, then, is part of any social order or historical period. Senses of movement include these historical senses of rhythm within them. Even the supposedly organic embodied rhythms of the walker vary historically: 'Old films show that our way of walking has altered over the course of our century: once jauntier, a rhythm that cannot be explained by the capturing of images'.[45]

Crucially, for Lefebvre, rhythm is implicated in the production and contestation of social order for 'objectively, for there to be change, a social group, a class or a caste must intervene by imprinting a rhythm on an era, be it through force or in an insinuating manner'.[46] Indeed, it is possible to see a particular politics of rhythm across a range of human activities. The rhythms of some kinds of music and dance, for instance, have famously upset those 'high up'. Jazz, punk and rave are but three examples of rhythms that have proved anxiety provoking to certain onlookers.[47] In the case of rave this led to the Criminal Justice Act of 1994 in the United Kingdom that explicitly referred to repetitive rhythms amongst its reasons for cracking down on people having fun. But rhythm is important in more sinister ways. 'Gait analysis' can now identify bodies moving with curious rhythms in airports and mark them for extensive searches and intensive surveillance. A strange rhythm of movements over a longer time period can similarly mark a person out. Too many one-way trips, journeys at irregular intervals, or sudden bursts of mobility can make someone suspect. Alongside these curious rhythms are the implicit correct and regular movements of the daily commute, the respectable dance or the regular movements of European business people through airports. There is an aesthetics of correct mobility that mixes with a politics of mobility.

Fourth – *What route does it take?* Mobility is channelled. It moves along routes and conduits often provided by conduits in space. It does not happen evenly over a continuous space like spilt water flowing over a tabletop. In Deleuze and Guatarri's account of nomadology they point out that it is not simply a case of free, mobile nomads challenging the 'royal science' of fixed division and classification. Mobility itself is 'channelled' into acceptable conduits. Smooth space is a field without conduits or channels. Producing order and predictability is not simply a matter of fixing in space but of channelling motion – of producing correct mobilities through the designation of routes.[48]

More concretely, Stephen Graham and Simon Marvin have developed the notion of a 'tunnelling effect' in the contemporary urban landscape.[49] They show how the routing of infrastructural elements ranging from roads to high-speed computer links warps the time-space of cities. Valued areas of the metropolis are targeted so that they are drawn into 'intense interaction with each other' while other areas are effectively disconnected from these routes.[50] Examples include the highways that pass though the landscape but only let you get off at major hubs. Or think of high speed train lines that pass from airport to city centre while bypassing the inner-city in between. These 'tunnels' facilitate speed for some while ensuring the slowness of those who are by-passed. Routes provide connectivity that in turn transforms topographical space into topological and, indeed, dromological space: 'Space-time no longer corresponds to Euclidean space. Distance is no longer the relevant variable in assessing accessibility. Connectivity (being in relation to) is added to, or even imposed upon, contiguity (being next to)'.[51]

Think of the development of a commuter rail network in Los Angeles. Built at huge expense to facilitate speedy transit from suburb to city centre it effectively by-passed the predominantly black and Hispanic areas of the city. While train riders were disproportionately white, bus riders were overwhelmingly black, Hispanic and female. A radical social movement, the Bus Riders Union, took the Metropolitan Transit Association (MTA) to court in order to halt the use of public money to fund the train system at the expense of the bus system. In court the MTA made the claim that train lines passed through many minority areas of the city such as Watts. In response, the Bus Riders Union argued that the population of areas the train lines passed through was not the relevant fact. The arrival of the train line had been matched by the removal of bus services. While the bus services had stopped frequently along the corridor (serving a 95 per cent minority community) the train hardly stopped at all and thus tended to serve white commuters travelling comparatively long distances. In addition, the BRU pointed out that the Blue Line was built at grade (rather than being underground or elevated), and had resulted in a high number of accidents and deaths in inner-city minority communities. So not only did the rail system produce 'tunnelling effects' by passing through minority areas it was also logically and economically related to a decrease in convenient bus routes and an increase in rates of death and injuries among inner city residents.[52]

Fifth – *How does it feel*? Human mobility, like place, surely has the notion of experience at its centre. Thus Bob Dylan's question 'How does it feel? To be on your own? With no direction home? Like a rolling stone?' is a pertinent one. Moving is an energy-consuming business. It can be hard work. It can also be a moment of luxury and pampering. The arrangement of seats on a trans-Atlantic flight is an almost perfect metaphor for an experiential politics of mobility. Upper, first or connoisseur class provides you with more space, nicer food, more oxygen, more toilets per person, massage, limousine service, media on line. Those at the back are cramped, uncomfortable, oxygen-starved and standing in line for the toilet. And then there might be the body, frozen and suffocated in the under-carriage well waiting to drop out in a suburb of a global city.

Consider walking once more. Tim Ingold has described how walking (and pretty much all manner of travelling) was experienced as drudgery and work by the well to do. 'The affluent did not undertake to travel for its own sake, however, or for the experience it might afford. Indeed the actual process of travel, especially on foot, was considered a drudge – literally a travail– that had to be endured for the sole purpose of reaching a destination'.[53] Before the Romantic poets turned walking into an experience of virtue 'Walking was for the poor, the criminal, the young, and above all, the ignorant [...]. Only in the nineteenth century, following the example set by Wordsworth and Coleridge, did people of leisure take to walking as an end in itself, beyond the confines of the landscaped garden or gallery'.[54] And even then the experience of walking was connected to the development of mechanized forms of transport that allowed the well to do to get to scenic environments for walking. Poor people, unaffected by the peripatetic poetry of Wordsworth and Coleridge, presumably did not experience walking in a new, more positive way. It was still drudgery.

Sixth – *When and how does it stop*? Or to put it another way – what kind of *friction* does the mobility experience? There is no perpetual motion machine and, despite the wilder prophecies of Virilio and others, things do stop. Spatial scientists famously formulated the notion of the 'friction of distance' as part of the development of gravity models.[55] Here it is the distance between two or more points that provides its own friction. But in a world of immediacy that is rarely flat and isotropic and where connectivity has become the most 'relevant variable in assessing accessibility', forms of friction are more particular and varied. As with the question of reasons for mobility (motive force) we need to pay attention to the process of stopping. Is stopping a choice or is it forced?

Graham and Marvin, in their consideration of a city of flows draw on the work of Castells and Ezachieli to point out that the new points of friction are not the city walls but newly strengthened local boundaries: 'global interconnections between highly valued spaces, via extremely capable infrastructure networks, are being combined with strengthening investment in security, access control, gates, walls, CCTV and the paradoxical reinforcement of local boundaries to movement and interaction within the city'.[56]

One of the effects of tunnelling is to produce new enclaves of immobility within the city.[57] Social and cultural kinetics means reconsidering borders. Borders, which once marked the edge of clearly defined territories are now popping up everywhere.[58] Airports are clearly borders in vertical space.

Often certain kinds of people, possibly those with suspicious rhythms, are frequently stopped at national borders. Sometimes for hours, sometimes only to be sent back. Black people in major cities across the west are still far more likely to be stopped by police due to racial profiling and the mythical crime of 'driving while black'.[59] In post 9/11 London, people of middle-eastern appearance are increasingly stopped by the police on suspicion of activities associated with terrorism. In the most extreme case, in July 2005 Jean-Charles de Menezes, a Brazilian man mistaken for a middle-eastern terrorist, was shot in the head seven times to stop him from moving on a London underground train. Racial profiling also appears to take place in airports in western nations where non-white people are frequently stopped and searched in customs or before boarding a flight. Those 'high up', meanwhile can stop and enjoy the scenery while others work at frantic pace around them. Friction is variably distributed in space and is an important component of mobility studies.

So here then we have six facets of mobility, each with a politics. The starting point, speed, rhythm, routing, experience and friction. Each is important in the creation of a modern mobile world. Each is linked to particular kinds of mobile subject identities (tourists, jet-setters, refugees, illegal immigrants, migrant labourers, academics) and mobile practices from walking to flying.

CONSTELLATIONS OF MOBILITY

So far I have outlined the importance of movement, representation and practice to the study of mobility. I have shown how each of these is implicated in the production and reproduction of power relations. In other words, how they are political. I have also suggested six facets of mobility that can serve to differentiate people and things into hierarchies of mobility. I would argue that each of these needs to be taken into account to provide accounts of mobilities at any given time. In the following section I develop a notion of *Constellations of Mobility* as a way of accounting for historical senses of movement that is attentive to movement, represented meaning and practice and the ways in which these are interrelated.

The ways in which physical movement, representations and mobile practices are interrelated vary historically. There is no space here for a charting of changing constellations of mobility through history. Rather I illustrate this point with reference to feudal Europe and its continuing influence on the contemporary mobile world with particular reference to the *regulation* of mobility. A key point of this section, and the chapter in general, is to dampen the enthusiasm for the 'new' that characterizes some of the work in the 'new mobilities paradigm', and to illustrate the continuation of the past in the present.

Carefully controlled physical movement characterized a feudal European sense of movement where the monopoly on the definition of legitimate movement rested with those at the top of a carefully controlled great chain of being. The vast majority of people had their movement controlled by the lords and the aristocracy. For the most part mobility was regulated at the local level. Yet still mobile subject positions existed outside of this chain of command in the minstrel, the vagabond, and the pilgrim. Within this constellation of mobility we can identify particular practices of mobility, representations of mobility and patterns of movement. In addition there are characteristic spaces of mobility and modes of control and regulation.[60] This was the era of frankpledge and of branding. As feudalism began to break down a larger class of mobile masterless men arose who threatened to undo the local control of mobility.[61] New subjects, new knowledges, representations and discourses and new practices of mobility combined. The almshouse, the prison and the work camp became spaces of regulation for mobility. By the nineteenth century in Europe the definition and control of legitimate movement had passed to the nation-state, the passport was on the horizon, national borders were fixed and enforced.[62] New forms of transport allowed movement over previously unthinkable scales in short periods of time. Narratives of mobility-as-liberty and mobility-as-progress accompanied notions of circulatory movement as healthy and moral.[63] By the twentieth century, mobility was right at the heart of what it is to be modern. Modern man, and increasingly modern woman, was mobile. New spaces of mobility from the boulevard to the railway station (the spaces of Benjamin's *Arcades Project*) became iconic for modernity. New subject positions such as tourist, citizen, globetrotter and hobo came into being.

Broadly speaking, the scale of regulation for mobility has moved in the past 500 years or so from the local to the global. While mobility of the poor was always a problem for those high up it was a more local problem in feudal Europe where wandering vagabonds were regulated by the local parish through a system known as frankpledge.[64] By the eighteenth century, mobility was beginning to become a national responsibility, passports were just around the corner and poor people moved over greater distances and more frequently. By the end of the nineteenth century the nation state had a monopoly on the means of legitimate movement and national borders, for the first time, became key points of friction in the movement of people.[65] By World War II, passports had become commonplace and nations were co-operating in identifying and regulating moving bodies. In each case it was indeed bodies that proved to be the key element even as the scale of mobility expanded and speeded up. While feudal vagabonds had their bodies branded like cattle, later travellers had to provide a photograph and personal details including 'distinguishing marks' for the new passes and passports that were being developed.[66] Now we are in a new phase of mobility regulation where the means of legitimate movement is increasingly in the hands of corporations and trans-national institutions. The United Nations and the European Union, for instance, have defined what counts and what does not account as appropriate movement. The Western Hemisphere Travel Initiative is seeking to regulate movement between the United States, Canada, Mexico and the Caribbean in ever more sophisticated ways.[67] Increasingly national interests are combined with so called pervasive commerce as innovative forms of identification based on a hybrid of biometrics and mobile technology are developed.[68]

One of the latest developments in mobile identification technology is the Rfid (Radio Frequency Identification) chip. These chips have been attached to objects of commerce since the 1980s. The Rfid chip contains a transponder that can emit a very low power signal that is readable by devices that are looking for them. The chip can include a large amount of data about the thing it is attached to. Rfid chips have the advantage over barcodes of being readable on the move, through paint, and other things that might obscure it, and at a distance. It is, in other words, designed for tracking on the move. Unlike a barcode it does not have to be stationary to be scanned. And Rfid technology is being used on people. As with most kinds of contemporary mobility regulation the testing ground seems to be airports. In Manchester airport a trial has just been conducted in which 50,000 passengers were tracked through the terminal using Rfid tags attached to boarding passes. The airport authorities have requested that this be implemented permanently. Washington State together with the Department of Homeland Security has recently conducted a trial involving Rfid tags on state driving-licenses allowing the users to travel between the states participating in the Western Hemisphere Travel Initiative. These tags can include much more information than is normally found on a driver's-license and can, of course, be tracked remotely.

It is experiments such as these that have led some to predict the development of a global network of Rfid receivers placed in key mobility nodes such as airports, seaports, highways, distribution centres, and warehouses, all of which are constantly reading, processing, and evaluating people's behaviours and purchases.

Information gathering and regulation such as this is starkly different from the mobility constellations of earlier periods. Regulation of mobility, to use Paul Virilio's term, is increasingly dromological. Dromology is the regulation of differing capacities to move. It concerns the power to stop and put into motion, to incarcerate and accelerate objects and people.[69] Virilio and others argue that previous architectural understandings of space-time regulation are increasingly redundant in the face of a new informational and computational landscape in which the mobility of people and things is tightly integrated with an infrastructure of software that is able to provide a motive force or increase friction at the touch of a button.[70] The model for this new mode of regulation is logistics. The spaces from which this mobility is produced are frequently the spatial arrangement of the database and spreadsheet.

CONCLUSION

The purpose of this chapter is to raise a series of questions about the new mobilities paradigm and to suggest some ways in which a mobilities approach can develop. I have suggested two caveats it is necessary to take on board in contemporary mobility research. One is an awareness of the mobilities of the past. Much that passes for mobilities research has a flavour of technophilia and the love of the new about it. In this formulation it is *now* that is mobile while the past was more fixed. We only have to consider the words of the 1909 Futurist Manifesto to see how this is a recycled notion. Consider point 8: 'We are on the extreme promontory of the centuries! What is the use of looking behind at the moment when we must open the mysterious shutters of the impossible? Time and Space died yesterday. We are already living in the absolute, since we have already created eternal, omnipresent speed'.[71] Here too the present and the future were about the mobile and the dynamic while the past was about stasis and stagnation. Yet their dynamism now seems quaintly nostalgic. Nothing seems more archaic than the futures of the past.

Taking an even longer backward look into history, consider the role of the medieval vagrant in the constitution of contemporary mobilities. It was the presence of these masterless men that prompted the invention of all kinds of new forms of surveillance and identity documentation that form the basis for what is going on now in airports and at national borders.[72] The figure of the vagabond, very much a mobile subject of fifteenth-century Europe, still moves through the patterns, representations and practices of mobility in the present day.[73] We cannot understand new mobilities, then, without understanding old mobilities. Thinking of mobilities in terms of constellations of movements, representations and practices helps us avoid historical amnesia when thinking about and with mobility. Reflecting Raymond Williams' notions of emerging, dominant and residual traditions that work to shape cultural formations we can think of constellations of mobility as emerging, dominant and residual.[74] Elements of the past exist in the present just as elements of the future surround us.

The second caveat is that in addition to being aware of continuities with the past that make contemporary mobilities intelligible we need to keep notions of fixity, stasis and immobility in mind. As proponents of the mobility turn have shown, mobilities need moorings.[75] Even the seemingly frictionless world of global capital needs relative 'permanences' in order to reproduce itself.[76] So while there is a temptation to think of a mobile world as something that replaces a world of fixities (Virilio's dromology is an example of this), we need to constantly consider the politics of obduracy, fixity and friction. The dromological exists alongside the topological and the topographical.

Finally, in addition to recognizing the importance of historical constellations of mobility in understanding the present and taking on board the importance of fixity I have argued that mobility itself can be fine-tuned through considering more specific aspects of mobility, each of which has its own politics and each of which is implicated, in different ways, in the constitution of kinetic hierarchies in particular times and places.

Note: This paper first appeared in *Environment and Planning D: Space and Society*, no 28 (2010), pp. 17-31. The editors are grateful to Pion Ltd. for the permission to reprint.

ENDNOTES

1   Kevin Hannam, Mimi Sheller and John Urry, 'Mobilities, immobilities and moorings', *Mobilities* 1/1 (2006), pp. 1–22; Mimi Sheller and John Urry, 'The New Mobilities Paradigm', *Environment and Planning A* 38/2 (2006), pp. 207–226.

2   Jørgen Ole Bærnholdt and Kirsten Simonsen, *Space Odysseys: Spatiality and Social Relations in the 21st Century* (Aldershot: Ashgate, 2004); Tim Cresswell, *On the Move: Mobility in the Modern Western World* (London: Routledge, 2006); Tim Cresswell and Peter Merriman (eds), *Geographies of Mobility: Practices, Spaces, Subjects* (London: Ashgate, 2008); Vincent Kaufmann, *Re-Thinking Mobility: Contemporary Sociology* (Aldershot: Ashgate, 2002); Mimi Sheller and John Urry, Mobile Technologies of the City (New York: Routledge, 2006); John Urry, *Sociology Beyond Societies: Mobilities for the Twenty-First Century* (London: Routledge, 2000); John Urry, *Mobilities* (Cambridge: Polity, 2007); Tanu Priya Uteng and Tim Cresswell (eds), *Gendered Mobilities* (Aldershot: Ashgate, 2008).

3   John Urry, 'The System of Automobility', *Theory, Culture and Society* 21/4–5 (2004), pp. 25–39; Peter Merriman, *Driving Spaces: A Cultural-Historical Geography of England's M1 Motorway* (Oxford: Blackwell, 2007); Jörg Beckmann, 'Automobility – a Social Problem and

Theoretical Concept', *Environment and Planning D: Society and Space* 19/5 (2001), pp. 593–607.

4   Peter Adey, 'Secured and Sorted Mobilities: Examples from the Airport', *Surveillance and Society*, 1/4 (2004), pp. 500–519 and, 'Surveillance at the Airport: Surveilling Mobility/Mobilising Surveillance', *Environment and Planning A* 36/8 (2004), pp. 1365–1380.

5   Alison Blunt, 'Cultural geographies of migration: mobility, transnationality and diaspora', *Progress in Human Geography* 31 (2007), pp. 684–694.

6   Urry, *Mobilities*; Cresswell, *On the Move*.

7   Bruno Latour, *Reassembling the Social: An Introduction to Actor-Network-Theory* (Oxford: Oxford University Press, 2005).

8   Thomas Kuhn, *The Structure of Scientific Revolutions* (Chicago: University of Chicago Press, 1996).

9   Georg Simmel, *The Sociology of Georg Simmel* (Glencoe, Illinois: Free Press, 1950); Robert Park and Ernest Burgess, *The City: Suggestions for Investigation of Human Behavior in the Urban Environment* (Chicago: University of Chicago Press, 1925).

10 Carl O. Sauer, *Agricultural Origins and Dispersals* (New York: American Geographical Society, 1952).

11 Ronald Abler, John Adams and Peter Gould, *Spatial Organization: The Geographer's View of the World* (New Jersey: Prentice Hall, 1971).

12 Pip Forer, 'A Place for Plastic Space?' *Progress in Human Geography* 2 (1978), pp. 230–267.

13 Susan Hanson and Geraldine Pratt, *Gender, Work, and Space* (London: Routledge, 1995); Laurie Pickup, 'Hard to Get Around: A Study of Women's Travel Mobility', in Jo Little, Linda Peake and Pat Richardson (eds), *Women in Cities: Gender and the Urban Environment* (New York: New York University Press, 1988).

14 Cresswell, *On the Move*.

15 This use of the term constellation reflects the use of the term by Walter Benjamin, *The Arcades Project* (Cambridge, Massachusetts: Belknap/Harvard University Press, 1999).

16 Robert D. Bullard, and Glenn S. Johnson, *Just Transportation: Dismantling Race and Class Barriers to Mobility* (Gabriola Island, BC: New Society Publishers, 1997); Brian S. Hoyle and Richard D. Knowles, *Modern Transport Geography* (New York: Wiley, 1998).

17 Brian Price, 'Frank and Lillian Gilbreth and the Manufacture and Marketing of Motion Study, 1908–1924', *Business and Economic History*, Second Series, 18 (1989), pp. 88–98; Ernest J. Yanarella and Herbert G. Reid, 'From 'Trained Gorilla' To 'Humanware': Repoliticizing the Body-Machine Complex between Fordism and Post-Fordism', in Theodore Schatzki and Wolfgang Natter (eds), *The Social and Political Body* (New York: Guilford, 1996).

18 Jonas Eliasson and Lars-Göran Mattson, 'It Is Time to Use Activity-Based Urban Transport Models? A Discussion of Planning Needs and Modelling Possibilities', *Annals of Regional Science* 39/4 (2005), pp. 767–789.

19 Adey, 'Secured and Sorted Mobilities'.

20 Allen White, 'Geographies of Asylum, Legal Knowledge and Legal Practices', *Political Geography* 21/8 (2002), pp. 1055–1073; Patricia Tuitt, *False Images : Law's Construction of the Refugee* (London: Pluto Press, 1996).

21 Nicholas K. Blomley, 'Mobility, empowerment and the rights revolution', *Political Geography* 13/5 (1994), pp. 407–422.

22 Rebecca Solnit, *Wanderlust: A History of Walking* (New York: Viking, 2000).

23 Tim Cresswell, 'The Production of Mobilities', *New Formations* 43 (Spring 2001), pp. 3–25.

24 Pierre Bourdieu, *The Logic of Practice* (Palo Alto: Stanford University Press, 1990).

25 Maurice Merleau-Ponty, *The Pheonomenology of Perception* (London: Routledge and Kegan Paul, 1962); David Seamon, *A Geography of the Lifeworld: Movement, Rest, and Encounter* (New York: St. Martin's Press, 1979).

26 Michel de Certeau, *The Practice of Everyday Life* (Berkeley, California: University of California Press, 1984), p. 101.

27 Marshall Berman, *All That Is Solid Melts into Air: The Experience of Modernity* (Harmondsworth: Penguin, 1988).

28 Keith Tester (ed.), *The Flaneur* (London: Routledge, 1994).

29 Nigel Thrift, 'Driving in the City', *Theory, Culture and Society* 21/4/5 (2004), pp. 41–59.

30 John Wylie, 'A Single Day's Walking: Narrating Self and Landscape on the South-West Coast Path', *Transactions of the Institute of British Geographers* 30/2 (2005), pp. 234–247; Tim Ingold, 'Culture on the Ground: The World Percieved through the Feet', *Journal of Material Culture* 9/3 (2004), pp. 315–340.

31 Ginette Verstraete, 'Technological Frontiers and the Politics of Mobility in the European Union', *New Formations* 43 (Spring 2001), pp. 26–43; Etienne Balibar, *We, the People of Europe? Reflections on Transnational Citizenship* (New Jersey: Princeton University Press, 2004).

32 Michael Oliver, *Understanding Disability: From Theory to Practice* (Basingstoke: Macmillan, 1996), p. 104.

33 Ernest Burgess archives of the University of Chicago Special Collections, box 126, p. 13

34 Zygmunt Bauman, *Globalization: The Human Consequences* (New York: Columbia University Press, 1998), pp. 86–87

35 Nicholas Blomley, *Law, Space and the Geographies of Power* (New York: Guilford, 1994).

36 Saskia Sassen, *Guests and Aliens* (New York: New Press/ Norton, 1999).

37 Stephen Kern, *The Culture of Time and Space 1880-1918* (Cambridge, Massachusetts: Harvard University Press, 1983); John Tomlinson, *The Culture of Speed: The Coming of Immediacy* (Los Angeles, California: Sage, 2007); Paul Virilio, *Speed and Politics: An Essay on Dromology* (Los Angeles, California: Semiotext(e), 1986).

38 Virilio, *Speed and Politics*, p. 142.

39 Nigel Thrift, 'Inhuman geographies: landscapes of speed, light and power', in Paul Cloke (ed.), *Writing the Rural: Five Cultural Geographies* (London: Paul Chapman, 1994); Tomlinson, *The Culture of Speed*.

40 Tomlinson, *The Culture of Speed*, p. 147.

41 Henri Lefebvre, *Rhythmanalysis: Space, Time, and Everyday Life* (London, New York: Continuum, 2004); Tom Mels, *Reanimating Places: A Geography of Rhythms* (Aldershot: Ashgate, 2004).

42 Mels, *Reanimating Places*; Seamon, *Geography*.

43 Lefebvre, *Rhythmanalysis*, p. 9.

44 Ibid., p. 8.

45 Ibid., p. 38.

46 Ibid., p. 14.

47 Cresswell, *On the Move*.

48 Gilles Deleuze and Felix Guattari, *A Thousand Plateaus: Capitalism and Schizophrenia* (Minneapolis, Minnesota: University of Minnesota Press, 1987).

49 Steve Graham and Simon Marvin, *Splintering Urbanism: Networked Infrastructures, Technological Mobilities and the Urban Condition* (London: Routledge, 2001).

50 Ibid., p. 201.

51 Offner quoted in Graham and Martin, *Splintering Urbanism*, p. 200.

52 Cresswell, *On the Move*.

53 Ingold, *Culture on the Ground*. p. 321.

54 Ibid., p. 322.

55 Andrew Cliff, Ronald L. Martin and Keith Ord, 'Evaluating the Friction of Distance Parameter in Gravity Models', *Regional Studies* 8/3–4 (1974), pp. 281–186.

56 Graham and Marvin, *Splintering Urbanism*, p. 206.

57 Bryan S. Turner, 'The Enclave Society: Towards a Sociology of Immobility', *European Journal of Social Theory* 10/2 (2007), pp. 287–304.

58 Chris Rumford, 'Theorising Borders', *European Journal of Social Theory* 9/2 (2006), pp. 155–169.

59 David A. Harris, 'Driving While Black and All Other Traffic Offenses: The Supreme Court and Pretextual Stops', *Journal of Criminal Law & Criminology* 87/2 (1997), pp. 544–582.

60 Valentin Groebner, *Who Are You? Identification, Deception, and Surveillance in Early Modern Europe* (New York: Zone Books, 2007).

61 A.L. Beier, *Masterless Men: The Vagrancy Problem in England 1560–1640* (London: Methuen, 1985).

62 John Torpey, *The Invention of the Passport: Surveillance, Citizenship, and the State* (Cambridge: Cambridge University Press, 2000).

63 Richard Sennett, *Flesh and Stone: The Body and the City in Western Civilization* (New York: Norton, 1994).

64 Robert A. Dodgshon, *The European Past: Social Evolution and Spatial Order* (Basingstoke: Macmillan Education, 1987).

65 Torpey, *The Invention of the Passport*.

66 Groebner, *Who Are You*.

67 Emily Gilbert, 'Leaky Borders and Solid Citizens: Governing Security, Prosperity and Quality of Life in a North American Partnership', *Antipode* 39/1 (2007), pp. 77–98; Matthew Sparke, 'A Neoliberal Nexus: Economy, Security and the Biopolitics of Citizenship on the Border', *Political Geography* 25/2 (2006), pp. 151–180.

68 Gillian Fuller, 'Perfect Match: Biometrics and Body Patterning in a Networked World', *Fibreculture Journal* 39/1 (2004).

69 Paul Virilio, *Speed and Politics*.

70 Nigel Thrift and Shaun French, 'The Automatic Production of Space', *Transactions of the Institute of British Geographers* 27/3 (2002), pp. 309–335; Martin Dodge and Rob Kitchin, 'Flying through Code/Space: The Real Virtuality of Air Travel', *Environment and Planning A* 36/2 (2004), pp. 195–211.

71 Filippo Tommaso Marinetti, *Manifesto of Futurism* (1909). Available from http://www.cscs.umich.edu/~crshalizi/T4 PM/futurist-manifesto.html.

72 Groebner, *Who Are You*; Zygmunt Bauman, *Legislators and Interpreters* (Cambridge: Polity Press, 1987).

73 Tim Cresswell, 'The Vagrant/Vagabond: The Curious Career of a Mobile Subject' in Cresswell and Merriman (eds), *Geographies of Mobility*.

74 Raymond Williams, *Marxism and Literature* (Oxford: Oxford University Press, 1977).

75 Hannam, Sheller, and Urry, 'Mobilities'.

76 David Harvey, *Justice, Nature and the Politics of Difference* (Oxford: Blackwell, 1996).

13   Curating Views:
The Norwegian Tourist Route Project

Janike Kampevold Larsen

A peculiar network of architecturally framed routes is unfolding in Norway. An expanded park with meandering paths and fixed viewpoints, the Tourist Route Project displays carefully curated portions of Norwegian nature as views.

With the pronounced aim of bringing car tourists to the countryside, the Norwegian Public Roads Administration initiated in 1994 a four-year trial project – the Tourism Project. By 1999 a fare more ambitious undertaking, the Tourist Route Project (TRP) was established, a collaboration between the Roads Administration and *Norsk Form*, an organization promoting Norwegian design and architecture. Their aim was to designate and develop eighteen stretches of road throughout Norway as official tourist routes or scenic byways, at the gross expense of 216 million Norwegian Kroner or 36 million Dollars. Several of these roads are finished, and the plan is to complete the last one by 2020. TRP's focus is not mainly on reconstruction or maintenance of these scenic byways, although upgrading and even restoration takes place to a certain degree, but more intensely on the architectural structures built along the routes. The TRP organizes architectural competitions for pullout points, lay-bys, view points, parking spaces, ferry quays and visitor centres, and the selection process is based on artistic, architectural and planning value. So far more than 60 projects have been built and many more are under way.[1]

Unique in Norwegian road history, the Tourist Route Project represents a curious and interesting blend of exhibition practices. While seemingly choreographing Norwegian nature, selecting exemplary sequences for motor-tourists to view, the project actually curates nature, architecture and the road itself, conceived within an economically motivated, national strategy. This chapter looks at the particular conceptions of nature underlying this display strategy, and asks what traditions for installing-in-nature the Tourist Route Project implicitly and explicitly endorses.

Norwegian roads are determined by topography, and the TRP takes advantage of the vistas and undulations created by earlier roads' negotiations with topography. Except in the very North, the overall structure of the country's road system has remained unaltered since 1846.[2] Until that time, roads had been developed solely to provide communication and transport between settlements in the country's gnarly, mountainous, fjord cut topography.

Many of the TRP roads are recently listed – the Geiranger road (1882-88), the Trollstigen road (1928–1936) and the Sognefjell road (1936–1938), as well as the Old Strynefjell Road (finished 1894) – and they were all built with the partial aim of easing travel for tourists.

The designated TRP routes run mainly through mountain passes and over mountain plateaus, along the fjords of Western Norway, and along the coast, as is the case with the Atlantic Ocean Road or the Lofoten Road. The eighteen routes are not interlinked, but spread out across regions already famed for their spectacular natural sceneries. The Norwegian Public Roads Administration has a long tradition of building roadside amenities, and simple rest areas have frequently been established next to natural viewing sites. The TRP, however, aims for a different 'process of cultivation', according to Karl Otto Ellefsen, member of the quality evaluation committee for the Tourist Routes.[3] The project cultivates already existing rest areas and creates new ones, securing and improving 'natural' and iconic viewing points like the Trollstigen Plateau, and the plateau by Vøringsfossen [the Vøring waterfall]. These have long been major tourist destinations and call for safer viewing situations. According to Ellefsen the intention at Vøringsfossen was to 'reinforce [...] the place's existing signature and narrative, without up-scaling the infrastructure to a point where the man-made landscape takes over'.[4] The winning entry, an elaborate project by Carl-Viggo Hølmebakk, features a pedestrian bridge across the river gorge, right above the point where the river bursts off the rocks into a 150 foot drop. On the brink of the main gorge there will be a visitor's centre, while secured footpaths will wind along the edges of the gorge, linking several viewing platforms.

In the TRP, the pragmatic and technically oriented Norwegian road building traditions have advanced towards a nationally orchestrated and politically informed aesthetics of nature as part of a built and choreographed environment. The TRP raises new questions as to how Norway conceives of its nature, and how landscape installations display, interpret, inform and mirror nature, as well as our experience of it.

The Tourist Route Project represents the first effort in Norway to include infrastructure in the national iconography. It does however compare to major infrastructural projects that established the basis for Norwegian industrialization, modernization and welfare systems. Ellefsen, for instance, draws parallels to channels and ports along the fierce Norwegian coastline.[5] More important still is the vast system of hydroelectric dams and plants exploiting the majority of Norwegian watersheds. Contrary to American examples like the Hoover Dam and the Norris Dam, most Norwegian dams were never designed to be visited, and remain practically invisible.[6] A second major engineering enterprise relevant to understanding the Tourist Routes Project is the Norwegian oil industry, starting with the construction of the world's first concrete oil storage tank, the Ekofisk Tank (1971–1973). The platforms moving out to sea are overwhelming sights as the platforms are still floating.[7] And yet, these visually spectacular achievements did not and still do not receive much public attention. The phlegmatic attitude with which the country has treated its mega structures, never including them in the national iconography, may seem surprising but has deep historical roots. Norwegians' relationship to nature has been a pragmatic one, a question not of conquest but of survival – of exploiting one's turf by modest means and providing a means of subsistence. It has been a question of self-protection as well as environmental protection, potent issues in relation to dam and power line construction.[8] As the TRP project is developing it suggests that the official Norwegian relation to nature is becoming invested with a new willingness to display that which Lewis Mumford once called man's relationship to land, and to show '[n]ature as a system of interests and activities'.[9]

DESCRIBING NORWAY

The Tourist Routes are mainly established on roads built between 1880 and 1940, an era when Norwegian views of nature, like North American ones, were moving from romantic idealization towards a more geographically and topographically informed approach. An important source for understanding this shift is the yearbook of the *Norwegian Tourist Association* established in 1868, with the extraordinarily dedicated road builder Hans Hagerup Krag as one of its founding members.[10] The yearbook's descriptions of Norwegian mountain routes and panoramas are clearly informed by the visual vocabulary of the late eighteenth century. Views are described as pictures, and the word 'panorama' is recurrent.[11] A picturesque vocabulary of the interesting, beautiful, arresting and sublime, is still predominant. We do however also find expressions like 'topographical plasticity' heralding a discourse on natural formations that fuses geographical exactitude with the individual aesthetic experience.[12] S. Høegh, for instance, expresses a typical rationalist attitude when he claims that 'nameless mountains are a heavy burden', and ventures to name and describe as much of the north-western mountain region as possible on his many trips in the region.[13] The yearbooks from this period feature numerous scientific articles on issues such as the depth of Norwegian fjords or the character of the country's surfaces, and news on road building make up a significant portion of the yearly reports. A particularly interesting example is an article in the 1881 yearbook by F.W. Heyerdahl, which promoted extensive road building in Jotunheimen, a famous mountain region in central Norway. This was necessary, Heyerdahl argued, if Norway wanted to challenge Switzerland as a prime destination for wealthy mountain tourists:

It is not only to its magnificent nature that Switzerland owes these many travelers, but certainly also to those accommodations with which one has equipped those places that travelers wish to go, in addition to the endeavors made to ease access to them.[14]

Among Norway's advantages, Heyerdahl listed the peculiar and extraordinary nature, the light and clean mountain air, the light summer nights and the possibility to roam freely. However, without infrastructure these virtues would never be discovered by tourists, maintained Heyerdahl, who considered road building the most certain way to bring 'considerable sums' into the country.[15]

Heyerdahl's argument corresponds closely to the overall aim of the Tourist Route Project: to encourage tourists to travel through Norway and thus strengthen the marginal economies of provincial settlements and towns. As the TRP official webpage states:

The country keeps losing ground in the tough international competition for tourists and struggles to be part of the global development. The challenge in developing the tourist routes is to create a tourist attraction that will receive national and international attention from vehicular tourists. The aim is to strengthen commercial interests and settlements, particularly in rural regions.[16]

Like Heyerdahl some 120 years before, the TRP argues that the roads should display the most spectacular regions of central and northern Norway. Heyerdahl celebrates 'the elevated and gripping beauty that our mountain plateaus with their cloud-roaring pinnacles and peaks display'.[17] The TRP claims that 'here, the travelers will experience Norway the way it is exhibited in tourist brochures and on postcards'. The Tourist Routes Project thus rests on a more than century-old endeavour to accommodate the tourist's movement and to direct her gaze.[18] Through advertisement, it seeks to re-mediate the Norwegian landscape, in a manner remarkably consistent with nineteenth-century landscape tourism. Much like America's main tourist route, the Blue Ridge Parkway, the Tourist Route Project aims at encouraging Norwegians to explore their own country.[19]

The exhibition *Detour* clearly demonstrates TRP's historical debt. Presenting all the major tourist routes, the exhibition has been touring Europe and North America since 2008, visiting 11 major cities including Washington, DC, and New York, before moving on to Shanghai in the spring of 2010, with Beijing and possibly Japan, Korea and India as future venues. Its main attraction is a looking cabinet showing live footage from various installations along the Tourist Routes in the different slots. The films have a fleeting, continuous, and very everyday-like character, and the display has a strong 3D effect. One sees people stop in their cars, get out, cross the road, look around, or keep driving – most of the films are shot in clear, dry weather. Accessibility is being stressed; everyone could visit these attractions.

The rotunda has a strong likeness to the *Kaiserpanorama* that was installed in 1883 in the 'Unter den Linden' passage in Berlin, and later reproduced in 250 German towns. This was not a true panorama, but a variation of it. The visual material was viewed from outside, not from its interior centre, and would typically be shown in sequences. An interior machine would rotate slides at a roughly two-minute interval, while the spectators peered through the lunettes covering the openings. Cycles of natural views were displayed featuring the inventor August Fuhrmann's vast collection of stereoscopic images from all over the world.[20] Jonathan Crary states that by 'employing transparent or color-tinted glass slides, Fuhrmann [...] provided some of the features (on a miniature scale) that were associated with Daguerre's original diorama: illusory, three-dimensional scenes whose reality effects were augmented by concealed lighting and translucent paintings'.[21] The irony of the *Detour* carousel is evident. This commercial device serves the exact opposite purpose from that of the original Kaiser-panorama: While the 1883 looking cabinet displayed landscapes and sceneries that were inaccessible to the public due to the cost of travelling, the Detour cabinet displays sites that are accessible; it serves as an incentive to motorized travel, to an ambulatory vacation that the majority of the public can actually afford. Film clips featuring regular people getting in and out of their regular cars demonstrate the democratic quality of the TRP. While the TRP as a commercial enterprise draws on historical viewing practices – not least on the distance between nature and viewer that these practices rest upon – most of the varied installations pretend to make nature accessible; pretend to bring us close. The installation's reliance on the panoramic tradition belies this attempt, however. By applying techniques from the historical panorama the exhibit maintains and increases the distance to nature inbuilt in the panoramic tradition. The TRP is about actual, not simulated travel, yet the project's highly mediated nature reflects and manipulates display strategies that are associated with the panorama, maintaining the panorama's distance to nature.

THE INSTALLATIONS

The various Tourist Routes Installations are very different in character. While choreographing Norwegian scenery, the installations in fact suggest a whole range of different approaches to nature, often referring explicitly to old measuring- and looking devices. One could argue that they serve as a dispersed looking cabinet to the European landscape tradition at large. Some of them are placed firmly within the panorama tradition, some refer to the early nineteenth-century picturesque, while still others echo earlier conceptions of nature as a sensuous field. They play with eighteenth and nineteenth-century tools and framing devices like the Claude glass, the panoramic platform, and the naturally elevated viewpoints from crags and bluffs that came into fashion with nineteenth-century landscape painting. Transcending any simple notion of facilitating presence, however, most of the installations are extremely conscious of their own form. They render it blatant that the project is about *mediation* – of scenery, of architecture, and of roads. In so doing, they mirror eighteenth-century representations of travel in Norway and the carefully constructed blend of artistry and national politics that Brita Brenna traces in the first chapter of this book. The TRP, similarly, is constructed through a combination of historical, aesthetic and economic considerations, displaying as distinct a 'feeling for nature' as did its nineteenth-century precursors. So what feeling for nature is at play in the Norwegian tourist route projects?

Several of the installations nurture an architectural language already familiar to city dwellers of the twenty-first century; cor-ten steel, glass, laminated wood and concrete are prevailing elements in the architecture of the rest

Figure 13.1 Staircase leading from Hellåga rest area to the Sjonafjord (photo by Jarle Wæhler)

National Tourist Route Helgeland North. Landscape architect: Landskapsfabrikken. Published with the permission of the Norwegian Public Roads Administration.

Figure 13.2 Detail from a telescopic viewing device at lower Oscarshaug (photo by Werner Harstad)

National Tourist Route Sognefjellet. The sighting device indicates the names of the surrounding mountains. Architect: Carl-Viggo Hølmebakk. Published with the permission of the Norwegian Public Roads Administration.

stops. By bringing architecture into nature, overstepping the gradation urban-rural-natural, the projects seem to subvert the very definition of a garden as an enclosure separated from second and first natures.[22] Interestingly, only one of the installations – the rest stop at Kjeksa along the Atlantic road – makes a point of this paradox and lays out a garden on the road shoulder.[23] This garden, however, is not enclosed and may be read as a symbol of the 'limitless' garden that nature is being turned into by the TRP.

In the 'garden' of the TRP there is a continuous negotiation between the panoramic, the close material views – and architecture. The low key rest area at Hellåga, for instance, employs a dual strategy, celebrating the view and disregarding it at the same time (Figure 13.1). At the upper level the view of the fjord dominates. The long corten stairs leading down to the lower level of soft slabs, however, seem to lead you straight into the sea. Attention is diverted from the panoramic view, and the downward focus towards the sea reveals instead the enticing materiality

at the site. The TRP contains several projects like these two – enhancing local places by discreetly accentuating existing features. The modesty of such projects is clearly at odds with the majority of the TRP installations, however, which tend to be less concerned about the landscape as sensuous field and more concerned with that one defining feature characterizing Western landscape conceptions since the late eighteenth century: the view. This tradition hypostatizes the position of the viewer and – by means of the central perspective – expels the subject from the landscape, now no longer the field we are *in* but a field that we *watch*. Our very gaze expels us from what we look at.[24] This dynamics is unfolded particularly clearly at Oscarshaug along the Sognefjell road, where Carl-Viggo Hølmebakk's viewing device bears a functional kinship to the Claude-glass, a tinted mirror used to frame and reflect the view (Figure 13.2).[25] Two vertical revolving glass sheets operate as a framing device and help the visitor aim at and thus isolate specific peaks for view, their location marked on the horizontal plane supporting the glass sheets.[26]

THE PANORAMA – PLAYING WITH DISTANCE

The majority of the TRP viewing platforms testify to the privileging of the view, so noticeable in the eighteenth-century garden, whether formal or picturesque.[27] The platforms mark the spot where the viewer should stand and watch nature – objectified and set up for view, much in the manner of the first plaques marking the perfect spot for perspectival viewing in eighteenth and early nineteenth-century picturesque parks. Even in the more modest installations this implicit viewing practice is noticeable, as in the small viewing platform at Askvågen.[28] A couple of the earliest installations, Sohlbergsplassen and Stegastein, are examples of elaborate viewing platforms turned into virtual landscapes.

Sohlbergsplassen, designed by the architect Carl-Viggo Hølmebakk, displays a highly mediated view, re-creating the outlook in Harald Sohlberg's iconic painting *Winternight in Rondane* (1914). The platform's interior modulation is highly metonymical, its poetically undulating concrete fences mimicking the shapes of the rounded Rondane mountains as well as invoking the undulating paths of a picturesque park. Saunders and Wilhelmsen's *Stegastein* on the other hand, is a large scale cantilevered wooden platform, a walking board running 33 meters off of the road shoulder into thin air. It carries attention more towards the vertiginous sensation of being there than to the expanded views ahead – not least due to the tilted glass banister at the end of the board. From a distance it seems to seamlessly connect the domesticated viewing platform and the wild nature outside of the 'garden'. As one gets near, however, one suddenly discovers the divide – and the precipice from which one is sheltered. This play with distance and sudden rupture gives it some of the same functions as the haha-ditch in the eighteenth-century garden.[29] A complex experience unfolds on and around this piece, as it metonymically invokes the roads that took you there – the road to nature so to speak.

A similar elaboration of the view is offered at Tungeneset on Senja, one of the northernmost installations of the TRP, where Code Architects have designed a rest area which includes an enclosed walkway functioning also as a viewing platform. The sidewalls of the heavy wooden construction are slanted and skewed – an irregularity possibly intended to pick up on the bulkiness of the slabs underneath. The

Figure 13.3 Sohlbergplassen rest stop (photo by Jørn Hagen and Werner Harstad)

National Tourist Route Rondane. View. Architect: Carl-Viggo Hølmebakk. Published the with permission of Norwegian Public Roads Administration.

Figure 13.4 Sohlbergplassen rest stop (photo by Jørn Hagen and Werner Harstad)

National Tourist Route Rondane. Detail. Architect: Carl-Viggo Hølmebakk. Published with the permission of Norwegian Public Roads Administration.

Figure 13.5 Stegastein viewing platform (photo by Janike Kampevold Larsen)

National Tourist Route Aurlandsfjellet. Architects: Saunders & Wilhelmsen.

photo featuring this platform on the TRP web pages shows an uncommonly pacific view of this extremely rough coastline.[30]

By elaborating the viewing platform, the TRP inverts main traits of the tradition of the panoramic rotunda. The panorama strived to obtain as full an illusion as possible, and this implied carefully articulating the area between the viewer and the painting. *Imitation terrain* signifies the praxis of actually imitating the grounds at the site, for example by replacing the actual platform with the poop deck of a frigate that had taken part in the battle, as the case was with Charles Langlois' rotunda in Paris, featuring the *Naval Battle of Navarino* (1827).[31] While the rotundas reproduced terrains that were inaccessible to the public at large, the TRP installations produce sleek architectural platforms in the *actual* terrain. While the panoramas sought to give viewers the feeling of being in the middle of the landscape or the actions depicted, the TRP installations strive to create a feeling of being *on a platform* in the middle of the landscape – by architectural mediation making the viewer aware that her relation to nature is staged. By thus inhabiting and furnishing the site

from which we look, the TRP perpetuates a significant trait in the panoramic tradition, while at the same time reversing it. They bring us close to the actual view, but maintain the distance that the panoramic rotundas of the late nineteenth century sought to eliminate.

TOPOGRAPHY AND GROUND

While the overall strategy of the TRP is one of display, there are a number of projects that pay attention to nature in very different ways. Opposed to the grandiose displays one finds differentiated close-ups on nature as materiality and diversity. Some projects are small-scale manipulations that encourage movement; a yellow handrail leading up a ridge to a promontory, or a footpath made of railway sleepers,

Figure 13.6 Viewing platform at Videseter (photo by JVA)

National Tourist Route Gamle Strynefjellsvegen. Architects: Jensen & Skodvin Architects. Published with permission of the architects.

both in Lofoten. Other projects ally themselves closely with the topology of the site, such as the ones designed by Jensen and Skodvin Architects, all providing close river views. The installations along the Old Strynefjell road display two small viewing platforms leading as close to the gorges and waterfalls as one can possibly come. These structures seem to seek neither grandiosity nor beauty, but base their construction instead on precise measurements of the terrain.

At Videseter, for instance, the shape of the fence is determined by following one datum line – a design strategy that stubbornly upholds topographical over aesthetic considerations, as the design is informed by the curving datum line. The viewing area Gudbrandsjuvet [Gudbrand's gorge] is more elaborate, consisting of two visiting areas displaying quite different architectural languages. One is an extensive access ramp, acutely angled in four stretches before it reaches the main viewing platform, an undulating and slightly slanted brim encircling a promontory and cantilevering the gorge. The installation brings the visitor in close proximity to the 'low matter' of forest floor; underbrush, birch and river. Yet the structure at large completely covers the portion of the terrain that forms the main viewing area, allowing surprising glimpses of the turbulent river underneath – filtered through striated grills.[32] Viewers are led straight into the landscape, invited to immerse in it – and still this dwelling is highly mediated.

NATURE ON DISPLAY

The TRP is about the dynamics between the individual experience and the staged aesthetics of the exhibition project. The installations, involving gorges, drops and waterfalls, are also about allowing the viewer to be safe while viewing. They thus depend on a convention that tells us that landscapes should be looked at from a distance. They rest upon distance as a productive principle, a convention for viewing which, as Gina Crandell shows, is many hundred centuries old. Tracing the conventions of pictorial perspective from fourteenth-century depictions of the annunciation, starting with Giotto's stage set paintings with a blue background illustrating the sky (cielo), Crandell shows how perspectivized landscape slowly evolved as backdrop to domestic spaces, ensuring 'the empowerment of the spectator'.[33] '[T]he depiction of distance between the horizon and the spectator provides the security of shelter and the feeling of control', she argues.[34] The TRP viewing platforms do indeed provide visitors with 'comfortable and secure position, buffered by a hefty space'. They are in fact taking this to the extreme, as many of the platforms are facilitating presence in 'impossible' places – they are cantilevering steep drops and colonizing sites that one could not otherwise safely access. They re-stage the intimate relation to the drop or the precipice that was so integral to Edmund Burke's definition of the sublime: The precipice induces the illusion of danger more than does heights. As one can actually fall into an abyss, it evokes the integral element in the sublime: '[T]he sublime is an idea belonging to self-preservation'.[35]

Although one can argue that Norwegian nature is not overtly sublime in character, it certainly holds a potential for sublime experiences – if one follows Burke's partial argument of the sublime belonging to self-preservation – as enhanced though the TRP viewing platforms. A number of them are cantilevering precipices, gorges and waterfalls – Stegastein's drop, Gudbrandjuvet hovering over the river gorge, the viewing platforms at Trollstigen plateau and the future platform at Vøringsfossen viewing area – they all exaggerate and hypostatize the precipice, placing us in 'impossible' situations where the architectural structure facilitates and intensifies experience of drop, height, danger. They create extreme conditions for viewing, while displaying nature.

ROADS ON DISPLAY

The Old Strynefjell road is a museum road per se, due to its retaining stone walls and gravel surface – which is also

Figure 13.7 Videdalen i Stryn with riksveg 258 (photo by Per Ritzler)
National Tourist Route Gamle Strynefjellsvegen. Published with permission of the Norwegian Public Roads Administration.

preserved. Of all the TRP projects, it is perhaps The Old Strynefjell road that most immaculately displays a *nature*. Unmediated by architectural installation, nature, here, is made accessible by the sculptural presence of the road itself. With its beautiful dry walls and soft curves, the old road takes us through the terrain in meandering curves – at the same time exposing and exhibiting the ground. The terrain is barren and desolate, but the distribution of rocks and snow renders its sheer, tinted materiality distinctive. The rocks are displayed as natural installations due to their

interchange with the road as an object. The neutrality of the rock-scattered plateau displays an unabashed singularity that prompts us to consider our being in the face of it.

The un-orchestrated road stretches provide access to a nature not visible from the grand tableaux of the tourist route installations. The sense of moving into the *it* of nature – that neutral senseless materiality of un-manipulated nature, may be the predecessor to the sublime experience of encountering abysmal gorges and steep mountains. These roads, as so many great drives, draw attention to the dynamics between open road and the specified viewpoint. More than the orchestrated tourist installations, they recall John Brinckerhoff Jackson's claim that we travel more for the sake of travel than to get to a specific destination.[36]

The grandeur of the 'regular views' of the TRP rests in their movement, in the shifting and sliding frames of the car window. While the installations' advanced framing devices create an illusion – a staged spectacle of nature – the fluctuation of views from the car window provides an individualized experience of the same area. As long as one is confined to motorized travel along these stretches, the car-window allows one at least the joy of movement, of seeing the 'screens' of the road shoulders and valley sides fold onto each other always forming new front screens as the road curves and undulates along them.[37] This joy of movement – of views rushing against and past us – reverses the relation between the spectacle and the spectator. Now, it is we who are being subjected to the views of the windshield, to the expanses of nature and the forms of infrastructure.

ENDNOTES

1   The Tourist Route Project involves 80 municipalities and 9 counties, as well as more than 20 tourist organizations, working closely with local tourist information groups, local administrations etc. While the overall structure – planning, selection of sites, the system of architectural installations – is a national responsibility, local administrations have a strong influence in planning and building the installations. More information is found on TRP's official website http://www.turistveg.no/.

2   North of Namsskogan near Trondheim Norway had no primary road network as late as 1914. Ref. Sverre Knutsen, *Veier til modernisering. Veibygging, samferdsel og samfunnsendring i Norge på 1800-tallet* (Oslo: Pax, 2009), p. 110. Several of the Tourist Route roads did not exist as more than riding paths at this point. The first

road across Sognefjellet was laid in 1915 by the telegraph company, see Herman Johan Foss Reimers, 'Over Sognefjed', in *Den norske turistforeningens årbok* (Oslo, 1922).

3   Karl Otto Ellefsen, 'Vøringsfossen: the remodulation of a tourist icon', *Magazine 'scape*, 1 (2010), p. 21.

4   Ibid., p. 21.

5   Ibid., p. 21.

6   An exception is the Sima power plant in Hardanger which receives large numbers of visitors each year. It was not planned as a tourist site, however, as were a number of US dams. At the Norris dam in Tennessee architect Roland Wand designed an '*architectural promenade* to lead visitors through all the key features of the dam, from the narrow crest overlooking the giant spillway and the downstream river, to the massive battered walls of the foot of the dam and the powerhouse of humming turbines', Christine Macy, Sarah Bonnemaison, *Architecture and Nature, Creating the American Landscape* (London: Routledge, 2003), p. 156. Another example is the much smaller Croton Dam in Westchester County, New York, part of the beautiful Croton Aqueduct system running from the Croton Reservoir that was supplying water for Manhattan between 1842 and 1940. This picturesque dam is equipped with a large English park right underneath it.

7   I am referring her to the sublime as a synthesis of Edmund Burke's and Kant's respective sublimes as introduced by David Nye as specific to technological installation in nature. See David Nye, *American Technological Sublime* (Cambridge, Massachusetts: MIT Press, 1994).

8   Twentieth-century dam extensions have met fierce resistance, the prime example being the resistance against development of the Alta-Kautokeino River, also involving the largest incident of civil disobedience in Norway. The development was commissioned in 1979 and criticized both for disrupting a watershed with unique biodiversity and for unsettling migrations paths for reindeer, thus threatening means of subsistence of

nomadic Sami populations. The hydroelectric power plant did not open until 1987.

9 Lewis Mumford, 'The Renewal of the Landscape', in *The Brown Decades, A Study of the Arts in America 1865–1895* (New York: Dover Publications, 1931), p. 26.

10 No: *Den norske turistforening*. Krag served as Norwegian Road Director from 1873 until 1903, and was responsible for most of the important mountain passes connecting the South-Easters part of Norway with the Western and North-Western parts of the country.

11 In the 1869 edition there is an article called 'Tre billeder fra Jotunheimen'/'Three pictures from Jotunheimen', and in the 1889 edition there is one called 'En storartet udsigt'/'A Magnificent View'.

12 Doctor S. Høegh in 'Udsigten fra Molde'/'The View from Molde', *Den Norske Turistforeningen Årbok* (Oslo, 1875), p. 91.

13 Ibid., 'navnløse fjeld ligger en tung for brystet', p. 92.

14 F.W. Heyerdahl: 'Det er dog ikke alene sin prægtige natur, Schweiz skylder denne masse reisende, man visselig også de bekvemmeligheder, hvormed man har vist at udstyre de steder, hvor de reisende så gjerne drager hen, samt de bestræbelser, som er gjort for at lette adkomsten dertil', *Den Norske Turistforeningens Årbok* (Kristiania, 1881), pp. 167–8.

15 Ibid., p. 168.

16 www.turisvegprosjektet.no. My translation.

17 Ibid., p. 169.

18 One of them was Keiser Wilhelm II who since 1869 brought his ship and entourage in several of the fjords in the North-Western part of Norway – to Geiranger, to Øye in the Nordangfjord, to Mundal in the Sognefjord.

19 See Thomas Zeller's chapter in this volume: 'Staging the Driving Experience: Parkways in Germany and the United States'. Zeller refers to the 'See America First'-movement.

20 Bernard Comment, *The Painted Panorama* (London: Harry N. Abrams Publishers, 1999), pp. 70–71.

21 Jonathan Crary, *Suspensions of Perception* (Cambridge, Massachusetts: MIT Press, 2001), p. 136.

22 Cf. John Dixon Hunt's classic differentiation between garden landscape, cultural landscape and wilderness by employing the terms first, second and third natures – where wilderness is the first. The limits were distinct in the seventeenth century and have gradually been permeated. Projects like the TRP seems to invert the logic even of the first and third nature, by installing 'gardens' in nature. John D. Hunt, *Greater Perfections, The Practice of Garden Theory* (London: Thames & Hudson, 2000), ch. 3.

23 Architect 3 RW – Jacob Røssvik, landscape architect – Arne Smedsvig.

24 At present we are completely immersed in a tradition of viewing objectified landscapes that started with the first perspectival landscape portrayals by Giotto in the sixteenth century. The perspective expels us. This issue is briefly discussed at a roundtable on landscape theory arranged and published by Rachel DeLue and James Elkins as *Landscape Theory* (London: Routledge, 2008).

25 The Claude Glass was used by looking in the small tinted mirror while turning one's back to the view. Thus the view would not only be formed and presented as in a picture, but it would be triply mediated. For a comprehensive account, see Arnault Maillet, *The Claude Glass: Use and Meaning of the Black Mirror in Western Art* (New York: Zone Books, 2009).

26 www.turistveg.no. Choose 'Sognefjellet', then 'Architecture'. Scroll down to 'Nedre Oscarshaug'.

27 The privilege of the view in Western pictorial history is the subject of Gina Crandell's *Nature Pictorialized, 'The View' in Landscape History* (Baltimore: Johns Hopkins University Press, 1993). It is also integral to picturesque studies of landscape design, particularly in William Gilpin's *Observations on the River Wye*, where the administration of the foreground, middle ground and background views in landscape painting is established as a landscape ideal, see *Observations on the River Wye*,

and several parts of South Wales relative to Picturesque Beauty (London, 1800). Claude Lorrain's painting are particularly influential in eighteenth-century English conceptions of landscape, also imprinting Humphrey Repton's and Uvedale Price's theories on picturesque garden arrangement, i.e. Price's *Essays on the Picturesque, as Compared with the Sublime and the Beautiful, and, on the Use of Studying Pictures for the Purpose of Improving Real Landscape* (London: Mawman 1810).

28 www.turistveg.no. Choose 'Atlanterhavsvegen', then 'Architecture', 'Askvågen' is the top project.

29 The sunken barrier used to keep live-stock out of gardens while avoiding interrupting the view to wilderness outside of the garden.

30 www.turistveg.no. Choose 'Senja', then 'Architecture', 'Tungeneset' is the top project on the page.

31 Comment, *The Painted Panorama*, p. 47.

32 Exhibition video can be viewed at www.jsa.no/exhibition.html.

33 Giotto, *Annunciation* (1305). Scrovengi Chapel, Padua. In Crandell, *Nature Pictorialized*, p. 60.

34 Ibid., p. 69.

35 Edmund Burke, *A Philosophical Enquiry into the Origin of our Ideas of the Sublime and Beautiful* (London: Oxford World Classics, 1968), p. 72.

36 J.B. Jackson: 'The Road Belongs in the Landscape', in Helen Lefkowitz Horowitz (ed.) *Landscape in Sight, Looking at America* (New Haven: Yale University Press, 1977), p. 253.

37 William Gilpin celebrated the spectacle of the folding screen in *Observations on the River Wye*: 'The views on the Wye, [...] are [...] exceedingly varied, [...] first, by the contrast of the screens: sometimes one of the side-screens is elevated, sometimes the other, and sometimes the front: or both the side-screens may be lofty, and the front either high or low. Again, they are varied by the folding of the side-screens over each other, and hiding more or less of the front. When none of the front is discovered, the folding side either winds round, like an amphitheatre, or it becomes a long reach of perspective'. Gilpin, *Observations*, p. 26.

14 Stop, Rest and Digest: Feeding People into Nature

Lars Frers

The route and the land through which it runs are in constant motion. They are not static entities over or through which dynamic men and their machines are moving. The route and the land are not taken out of time, they are not frozen on a canvas. As part of the earth's surface they revolve through day and night, they pass below clouds and showers of rain, they move from winter to spring, they become brittle and crack, are overgrown, maintained, and painted. In this chapter, I want to complement the perspective of a moving observer who develops judgments about the aesthetics of route and landscape by adding two new angles: first, the perspective of the dynamic route, second, the perspective of the corporeal, living human body. Of course, neither roads nor bodies have a perspective of their own. The idea is to take the actual entities that are relevant in this volume – routes and the land – as starting points that are of equal analytic value as constructs or concepts such as the sublime, landscape, or mobility. The idea is to examine what kind of agency these entities unfold in interactions with those who are *en route*. This interest in the dynamics that arise when different kinds of agency (natural, conceptual, technological, social, psychological, etc.) mix and interact is inspired by work in science and technology studies, where the division between human and non-human agency is not taken for granted.[1] Inspecting the ways in which the non-intentional/material and the intentional/social blend into each other, my analysis will situate itself in a peculiar place along the route or in the land – the rest stop or view point – and study what happens there.

FEEDING THE ROUTE AND BEING FED

The road itself is reserved exclusively for moving vehicles – stopping, being on the road without moving, is anathema to its regime. Places in which stopping is acceptable or even required have to be marked and highlighted in an attention-grabbing manner. There are several kinds of places that allow such stops, places which are designed and used in very specific ways, and riddled with regulations.[2] Points of particular interest such as remarkable views are marked up by means of signs and landscaping strategies, or indicated on maps and other navigation devices. Points of consumption enable the traveller and his/her vehicle to continue the trip, supplying food, drink and fuel, together with a range of other necessities such as souvenirs, fishing rods, skiing equipment, sun cream, condoms etc. Since people continuously digest and since they regularly produce waste while on the road, they need places to get rid of the digested food, empty packaging, left-overs and other unwanted stuff: points of excretion. Toilets and facilities for collecting different kinds of waste are announced along many touristic routes, and the need to use these facilities is often the most urgent reason for people to interrupt their automobility.

These three kinds of places participate in the production of the route as a very specific setting. Accordingly, the statement that stopping is anathema to the route has to be qualified: Without occasional stopping, the movement of humans and their vehicles cannot be sustained. In this chapter I will focus on these stopping places as part of the route. They are the site from which I can display a different perspective on movement and mobility. Inspired by Lefebvre, this chapter treats the production of the route as process based on a triad consisting of (a) its dynamic materiality, (b) ideas of the route, where it leads to and what it should look like, and (c) the lived experience of corporal beings *en route*.[3] The term process has a decidedly neutral tone, but the processes happening here are not neutral. The encounters of different entities and the crossing of different courses are often conflict-laden and rough. It could be understood as a mangle that forcefully blends different agencies, bringing forth something new and different.[4] However, seen from a less post-humanist and more sociological point of view, the processes at stake here can also be understood as performances. They are performative in a double sense: firstly, they are enacted on a certain stage and in a specific setting, but still open to contingencies, to sudden changes, disruptions and redirections. Secondly, they are performative in the sense that they are displayed, shown and made sensible to others who are co-present in the situation. The important point for the present work is to focus on the encounter, the active mingling and mangling of different kinds of agencies in a way that is perceived by others, presented to others and influenced by co-present others – be they human or non-human.

CO-MANAGING THE DISGUSTING AND THE SUBLIME

The two main locations that I have studied are part of the Norwegian Tourist Route project.[5] In her contribution to this volume, Janike Kampevold Larsen provides an in-depth analysis of the aesthetic ideals informing these installations. In this chapter, I want to supplement Larsen's account with considerations about aesthetics as *aisthesis*, i.e. as concerned with sensual perception. I will focus on how concepts such as the beautiful and the sublime enter the settings and how they mix and interact with the everyday perceptions and practices that are enacted at the sites.

The two sites chosen for this study vary in their material setup and in the way they present the landscape – spectacularly in case of the vertigo-inducing Stegastein platform, and mundanely or even ironically in the case of Hereiane's bright yellow toilets. Both sites are within a day's driving distance from the principal cities Oslo, Bergen, and Stavanger and lie close to popular routes that direct road traffic through Norway. My fieldwork was carried out in two periods: the first at the end of March, and the second at the end of June, during the warmest week of the year. In addition to my logged observations, photographs and video-recordings, I relied on one main methodical device to get access to the perceptions of those who spent time at these places: video narrations.[6] The intention was to get access to the ways in which the surroundings appear in someone's perception and to the ways in which someone moves her or his perception onto something. Put differently, this method helps to analyse what Waldenfels in his phenomenology of attention calls the tension between paying attention to something (*aufmerken*) and having one's attention drawn to something (*auffallen*).[7] Having the moving images of a film accessible for the analysis in synchronicity with the narration of the one who is filming is of course not the same as 'reading the mind' (which, from a phenomenology of the body's point of view, would necessitate 'reading the senses' too) of someone who lets his senses wander through his surroundings, but it is definitely much richer than simply observing someone or asking them to film and later talk about it (as in photo or video elicitation interviews). To focus the video narrations, I attached a small printout with three questions to the point-and-shoot camera and asked the respondents to answer the questions while filming:[8]

1. How did it come to your stop here?
2. Please film and talk about what is characteristic for this place. Take your time.
3. Is this place as you thought it would be?

Before analysing the responses, a brief introduction to the two sites will be useful. The following two photographs show the sites' main architectural features (the platform and the toilet building) in March, outside of the tourist season.[9] Taken off-season, both photos show the sites in a somewhat problematic state. The wet snow at Stegastein enters the encounter in a way that requires significant attention from those who want to walk along the platform. The existing tracks of downtrodden snow, material trace of former social practices, lead the way along the centre of the platform (Figure 14.1).

Driving off the road and entering the Hereiane rest stop, one quickly notices that the toilets are not open. Wooden planks have been nailed into the entrances and closer inspection reveals that the yellow glass doors have

Figure 14.1 Stegastein viewing platform in March (photo by Lars Frers)
National Tourist Route Aurlandsfjellet. Architects: Saunders & Wilhelmsen. @Lars Frers Creative Commons License by-nc-sa 3.0.

Figure 14.2 Hereiane toilet building in March (photo by Lars Frers)
National Tourist Route Hardanger. Architect: Asplan Viak/3RW. @Lars Frers
Creative Commons License by-nc-sa 3.0.

been destroyed. If the aim of the stop was to get rid of one's excrements in a civilized manner, then encountering the closed facilities can be a frustrating experience. Walking off into the landscape to search for a hidden spot to relieve oneself is one solution for this problem. The main difficulty of this solution is the management of visibility or perceptibility. The toilet building hides the person who urinates and/or defecates both visually and relating to sound and smell, whereas the relatively open landscape at this location only does so if one makes a substantial effort. It is necessary to look for an inconspicuous, safe and fast route which takes one out of the reach of the senses of others. In comparison to other rest stops, however, very little suspicious litter could be found in the vicinity, it being off-season. The toilets at Stegastein were closed too, without any visible reason or explanation other than the fact that freezing night temperatures might have forbidden the use of running water in an unheated building.

During the main season, the toilets at both locations were open. At this time of year they emerged as highly relevant actants in both settings, although in almost opposite ways.[10] At Stegastein, the main architectural sight is the platform. The rest of the architectural setting – the concrete benches, the waist-high walls built from roughly cut stone and the black-painted, concrete and wood toilet building – contributed to the presentation of the location but in a less remarkable way. To some degree, the narratives of the respondents were steered to the platform through my actions because I often approached them at the end of the platform, after they returned to its base. Nonetheless, the platform is *the* obvious main attraction of the site, and people spend most of their time there. Even when they are at other locations of the site they often refer to the platform with gestures and gazes. But many people still spent a substantial amount of time at the less spectacular part of this site. During the busy hours (from about ten o'clock to about seven o'clock in the afternoon) visitors often had to wait before they could enter the toilet. People perceived this waiting time to be longer than it should have been because one of the toilets, the one marked as the women's toilet, was permanently locked. What made things worse was that the men's toilet room also housed a urinal which was clogged and (for several days in a row) full of smelly, yellow urine. In two interviews, people remarked on the contrast between the wonderful site and the architecture,

Figure 14.3 Inside the Hereiane toilet (photo by Lars Frers)
Architect Asplan Viak/3RW. @Lars Frers Creative Commons License by-nc-sa 3.0.

which was characterized as probably very expensive, one the one hand and the poor state of the toilets on the other hand. Since the toilets are not the most salient topic it is remarkable that they were mentioned during the few minutes of a talk with a stranger.[11]

At Hereiane, the toilet building is the main architectural feature – the extravagant building even acquired the nickname 'the million dollar toilet' [milliondo] in the local press.[12] In the interviews and narrations, the toilet was repeatedly described as special, surprising, clean and/or beautiful.

Transcript 1[13]

W: [*reads instructions while walking towards the toilet building, keeping the building in the center of the image*] Please film and talk about what is characteristic for this place. Take your time. (1) .hhhh Well the <u>toilet</u> (.) is very <u>special</u> ehehehehehe |.khh hh |beauti(.)ful building
M: |yes it's an unusual |building yeah

Transcript 2

It wasn't really the <u>rocks</u> so much that a that <u>grabbed</u> my attention it was it was <u>this</u> [*centers the view on the toilet building*] (0.5) .hh ahm (0.5) I guess this facility here (.) .hh aahm which is pretty unusual a architecture

Several people remarked on the attractiveness not only of the exterior of the building but also talked about the interior facilities as being clean and unexpectedly pleasant (although no one filmed the interior). There are two separate toilets, without any signs regarding gender or disability status on the doors.[14] As can be seen in Figure 14.3, the interior contrasts with the exterior insofar as the slate stone surface is not flat and polished but rather rough-cut. The toilet, the sink and the trashcan along with a few other implements are made of either brushed or polished stainless steel. The floor and wall that separates the two toilets are painted in the same bright yellow colour as the outside floor, and the glass doors feature a semi-opaque, bright yellow plastic layer. One of the toilet rooms also has facilities to make it more accessible. The interior was warm, and usually it did not have a particularly unpleasant smell.

While the Stegastein toilet was used by those who stopped at the site, it was never characterized as inviting or pleasant. The smell and the sight of the clogged urinal had a serious impact on the place's feel. The Hereiane toilet, in contrast, was not only used by those who stopped there. Campers who stopped at this site overnight told me that the toilets (and access to running water) were one of the principal reasons why they decided to stay at this site for an extended period of time. These parties even tended to the toilets to some degree, picking up the odd piece of litter that short time visitors left on the floor in the building.

The ethnographic data on which this chapter is based is of course not sufficient to claim that there is a significant correlation between the design and the state of maintenance and functionality of a toilet on the one hand and the tourist experience as a whole on the other hand. However, the fact that the toilets emerged as a topic in the narratives and conversations at both sites points to the fact that they were perceived and treated as one of the main components of the site – a component which became an actant that elicited complaints (in the case of the Stegastein site) or praise (in the case of the Hereiane site). The toilets became part of the narrative about the experience of the site itself but they also featured as part of narratives about the touristic experience of Norway as a whole. In this narrative, the Hereiane toilets were seen as exemplary and very different to rest stop toilets in other places, either at home or in other touristic regions.

The need to excrete and urinate is part of the mobile experience. It goes along with the need for managing a taboo, and with managing one's temporary disappearance

Figure 14.4 View from Stegastein viewing platform (photo by Lars Frers)

@Lars Frers Creative Commons License by-nc-sa 3.0.

from the attention of others. Performing these urgencies, people are required to pay attention to their relation to others, to their own bodily processes, to concepts of hygiene, and to the perceivable traces left by other human bodies. These performances have a pleasant side – as the act of excretion, the feeling of relief and also the temporary solitude can be pleasurable – but they are also attached to a whole set of potentially unpleasant, disgusting or even painful experiences.

The two rest stops display unusual and even spectacular architecture, but they are also embedded in a geographic region that is typically Norwegian: the fjord. Renowned for their breathtaking views, where the sheer scale, the non-human forces at work, and the configuration of water, mountains and plant life creates sublime experiences, the

Figure 14.5 View from Hereiane (photo by Lars Frers)

@Lars Frers Creative Commons License by-nc-sa 3.0

fjords and the views they afford appeared in almost all of the narrations. Connections to the aesthetic ideal of the sublime were stronger at the Stegastein site, where the combination of a deep, vertigo-inducing descent to the Aurlandsfjord, the steep mountainsides and the waters of the fjord were experienced and described more often in terms connecting it to the idea of the sublime.[15]

Transcript 3

yeah it's actually more breathtaking than (.) than we ever thought it would be

The characterization of the aesthetic experience at Hereiane was somewhat closer to the beautiful, but it still contained references to a sublime dimension, to vastness[16] – although this vastness often also interpreted as being peaceful rather than awe-inspiring.

The following transcript follows on from the conversation about the toilet from the first transcript:

Transcript 4

M: [*sniff-like inhalation*] (1) but what is really
    special is <u>this</u> [1 sec, *camera moves to the
    right to show the fjord*]
W: the <u>view</u> hh
M: the view (.) and the <u>rocks</u> (.) they look very
    like very much like ehm (.) the rock formations
    in Australia (1)
W: .hh and its very peaceful |here
M: |and its very peaceful
    °here yes°
    [5 sec, *walking closer to the edge of the parking
    lot, trees coming into view*]
M: and beautiful (0.5) trees (2.5) they're <u>pines</u> .hh
    but they're t|iny
W: |yeah (1).hh everything is |tiny here

This sequence displays how quickly and easily the attention can shift from the toilet building to the surrounding landscape. In this transcript, the Australian man (M) also marks his contestation of the prior statement of his partner, a Swiss woman (W) who said in transcript 1 that the toilet building is special, by starting his utterance with an audible inhalation followed by a 'but', further emphasizing the weight of his claim by the use of the word really in 'what is really special' and emphasizing the contrast by stressing the '*this*'. In the next turn, his partner offers 'the *view*' as a suggestion for what could be special. The suggestion is then briefly acknowledged in the following turn, but particular stress is put on another aspect 'the *rocks*'. According to descriptions of Hereiane on information signs and in brochures, the rock formation, and the way it smoothly glides into the Hardangerfjord, is the main geographic feature that makes the landscape at Hereiane special. As can be seen in Figure 14.5, the gentle descent of the rock surface allows an easy transition from the parking lot to the rocks surrounding the site and then further down into the fjord (several people went swimming at this site and some were fishing too). Quick switches likes this, from one aspect of the site to another, from toilets to landscape, from road to benches, can be found in many of the video narratives. Sometimes the image shows a feature first and then the narration turns to that feature, but other times the narration turns to another aspect and then the camera is quickly moved to focus on the new topic too.

The narrations demonstrate that the proximity of mundane or even disgusting features or experiences to sublime or beautiful features or experiences is not uncommon, and that people have a ready repertoire of interactional and narrative tactics to address these transitions. Sometimes the contrast is accompanied by laughter or irony, sometimes in- or exhalations mark the shift, other times the only thing in-between is a slightly elongated pause. As the video narrations show, the spatial proximity and sometimes even the architectural or material proximity goes in parallel with this narrative proximity. Accordingly, it can be said that the sublime and the mundane are not mutually exclusive in the sense that they can only happen at very different sites or that a long time needs to pass to be able to have one kind of experience follow the other. On the contrary, the different agencies of the body, the material surroundings, and social or aesthetic concepts clearly overlap with each other. While this overlap can become problematic and require involved interactional responses to manage the transition from one activity or experience or topic to another, the same overlap is also used as an opportunity for further interactions, be they the making or sharing of complaints or for shared amusement.

AESTHETIC AGENCIES

When the route is understood as a process in which the different agencies of people, of material spaces, and of social or mental concepts are mangled then Tim Edensor's claim that mobility is as much about the mundane as it is about the extraordinary shows its full potency.[17] The rest stops of the Norwegian Tourist Route with their contrasting aesthetics – where nature and architecture, the eternal and the contemporary, the sublime and the disgusting are in close proximity to each other – unfold a very specific set of aesthetic agencies. People bring some of the agencies with them. They have ideas about what the place that they will stop at should look like. They follow routines, some dictated by the route itself, others dictated by social conventions and personal preferences. But in their perceptional practices they encounter the places – with their specific material agencies, the weather interacting with the land around them,[18] the current state of the facilities, the view that unfolds in their senses and in their cameras – as entities that all come together in the perception or *aisthesis* of these places. But more than that: As the world is not waiting for things to happen, people also are not passive recipients of external cues. They perform their own aesthetics. They position themselves in certain

places to produce a view. They manage their absence in the eyes of others when they go to the toilet and when they wait for someone to finish taking a photograph. They adjust their hair and their clothes, smile into a camera and crack jokes. Or they search for a spot where they can stand still for a while, immersing their senses in the local flow of time, embedding themselves into the landscape. In all of these performances, they display their expectations and their taste.[19] In these performances, they place themselves and the site on a field of aesthetic agencies that is open to shifts and movements, but that is also deeply hierarchical. To paraphrase some statements: 'You think the toilet is special? Look at the rocks!' 'This is maybe the most beautiful place in the world.' 'I have been here before, this is what I expected Norway to look like.' 'If only they would properly manage this site.'

But in my material it also becomes obvious how people display themselves, their taste and their perception of the place in their bodily performances, in playfulness, in furrowed brows, long looks or deep sighs. Thus, social and aesthetic hierarchies are put into play at these sites. In this article, I have focused on aesthetic categories – but other categories and identities are also negotiated in the context of these places: they impact perceptions of a nation state as a whole, à la 'Norway is a beautiful country', of economic developments and tourism 'they spent a lot of money on this building, maybe they should rather have fixed the holes in the road', and of one's own role in the world 'I can really let it all go here and think about my life'. The rest stop and the nation, the social and the individual – they all are in constant motion, negotiated by all the things and people who encounter each other at these sites.

ENDNOTES

1   Bruno Latour, *Science in Action: How to Follow Engineers and Scientists Through Society* (Cambridge, Massachusetts: Harvard University Press, 1987). Andrew

Pickering, *The Mangle of Practice: Time, Agency & Science* (Chicago: University of Chicago Press, 1995).

2   Peter Merriman, 'Materiality, Subjectification, and Government: The Geographies of Britain's Motorway Code', *Environment and Planning D: Society and Space* 23/2 (2005), pp. 235–250.

3   Henri Lefebvre , *The Production of Space,* trans. Donald Nicholson-Smith (Oxford: Blackwell, 1991), pp. 38–40.

4   Pickering, *The Mangle of Practice*; Andrew Pickering, 'Practice and Posthumanism: Social Theory and a History of Agency', in Theodore R. Schatzki, Karin Knorr-Cetina, and Eike von Savigny (eds), *The Practice Turn in Contemporary Theory* (London: Routledge, 2001), pp. 163–174.

5   More information and images of the different sites can be found on TRPs website: http://www.turistveg.no.

6   Previously explored in Lars Frers, 'Perception, Aesthetics, and Envelopment: Encountering Space and Materiality', in Lars Frers and Lars Meier (eds), *Encountering Urban Places: Visual and Material Performances in the City* (Aldershot: Ashgate, 2007), pp. 25–45; Lars Frers, 'Video Research in the Open: Encounters Involving the Researcher-Camera', in Ulrike Tikvah Kissmann (ed.), *Video Interaction Analysis: Methods and Methodology* (Frankfurt am Main: Peter Lang, 2009), pp. 155–177.

7   Bernhard Waldenfels, *Phänomenologie der Aufmerksamkeit* (Frankfurt am Main: Suhrkamp, 2004).

8   I provided these questions in Norwegian and German translation to offer at least one language in which the respondents were sufficiently fluent to understand the questions.

9   Both Figures 14.1 and 14.2 have been taken with a wide angle lens to capture more of the surroundings – with the effect that the objects in the centre of the image appear relatively smaller/more distant in comparison to their surroundings (this is also true of Figure 14.3).

10   Bruno Latour, *Reassembling the Social: An Introduction to Actor-Network-Theory* (Oxford: Oxford University Press, 2005).

11   Mostly because of their location at the site and because of the taboo quality of talking about defecation. To a certain degree the taboo is not very strong in this setting, because 'complaints' seemed to be a salient topic for a talk with a researcher that was perceived to be in some connection with those who are responsible for the state of affairs and who could forward their complaints. See also Diana Boxer, 'Social Distance and Speech Behavior: The Case of Indirect Complaints', *Journal of Pragmatics* 19/2 (1993), pp. 103–125 on complaints as a topic for conversations with strangers. I sometimes made this connection even more plausible by telling people that the project in which I am involved is also collaborating with the Norwegian Public Roads Administration (*Statens vegvesen*).

12   Hardanger Folkeblad, *Nasjonal Turistveg Jondal – Utne*, (2007, retrieved 08–11 2010), http://www.hardanger-folkeblad.no/hfkultur/article2832969.ece.

13   I am using standard conversation analytic transcription markup. See J. Maxwell Atkinson and John Heritage (eds), *Structures of Social Action: Studies in Conversation Analysis* (Cambridge: Cambridge University Press, 1984), pp. 9–16. Oblique text in square brackets gives information about what is going on, numbers in brackets indicate the length of pauses in seconds, full stops in brackets indicate a very short but audible pause, '.hh' and 'hh' indicates an audible in or exhalation, 'ehehe' indicates laughter, underline text indicates an emphasis, and pipe symbols '|' indicate the starting point of overlapping talk.

14   This has obviously changed. The photographs in Hardanger Folkeblad, *Nasjonal Turistveg Jondal – Utne* show that one toilet originally had a disability symbol on the door, while the other displayed a man and a woman symbol.

15   Immanuel Kant, *Kritik der Urteilskraft: Beilage: Erste Einleitung in die 'Kritik der Urteilskraft'* (Hamburg:

Meiner, 2006); Edmund Burke, *A Philosophical Enquiry Into the Origin of Our Ideas of the Sublime and Beautiful* (Oxford: Oxford University Press, 1990).

16 Uvedale Price, *Essays on the Picturesque, as Compared With the Sublime and the Beautiful: And on the Use of Studying Pictures for the Purpose of Improving Real Landscape*, 3 vols (Farnborough: Gregg International, 1971).

17 Edensor, 'Mundane Mobilities, Performances and Spaces of Tourism', *Social & Cultural Geography* 8/2 (2007), pp. 199–215.

18 Tim Ingold, 'Bindings Against Boundaries: Entanglements of Life in an Open World', *Environment and Planning A* 40/8 (2008), pp. 1796–1810.

19 Pierre Bourdieu, *Distinction: A Social Critique of the Judgement of Taste*, trans. Richard Nice (Cambridge, Massachusetts: Harvard University Press, 1984).

15 Roadside Aesthetics: Guidelines from the Norwegian Public Roads Administration

Beate Elvebakk

A sizeable part of Norway's road network passes through dramatic natural landscapes. Compared to other European countries, Norwegian roads may seem to exist in a closer relationship to nature, and to present the driver with pure and unspoilt landscapes. The landscapes observed from the road, however, are always mediated. Although the road is probably the most common place from which to observe the Norwegian landscape, shaping our notions of what Norway 'looks like', the landscape seen from the road is constructed in the most concrete sense of the word. Roads form new landscapes, they physically shape the terrain they run through, and, although this is not something that the traveller is usually aware of, the road is also often planned for its effect on the observer. The landscapes of roads are not 'natural' landscapes; they are, in a sense, non-existent landscapes – they represent a point of view that must always be made, and testify to an aesthetics derived from international conventions as well as local and national ideas and sensibilities about what is beautiful, meaningful, and worth seeing.

THE PLANNING OF BEAUTIFUL ROADS

While landscape architects have long been involved in the planning of roads in countries like Germany, the United States and United Kingdom,[1] the first landscape architect was employed by the Norwegian Public Roads Administration (NPRA) only in 1975.[2] Today, however, the organization employs between 70 and 80 landscape architects, and from 1992 every regional office is required to possess 'aesthetic competence'.

The landscape architects are not necessarily involved in every case of road construction or reconstruction, but for major construction or reconstruction works, a set of guidelines, variously referred to as aesthetic guidelines, visual guidelines, or guidelines for design, is usually developed as part of the planning process, typically in the form of a glossy brochure containing a mixture of description of current status and measures planned, maps, technical drawings, and photographs. This chapter is based on a selection of such guidelines published in Norway over the last fifteen years,[3] along with a selection of other documents on road aesthetics published by the NPRA. It will try to present on the one hand the landscape, so to speak, in which these aesthetic judgments take place, and, on the other hand, the idea of the road and the landscape as aesthetic objects or spheres that emerge from the documents.[4]

The aesthetic guidelines relate to projects that are relatively major roads in a Norwegian context, but mostly fairly modest by international standards, with two to four lanes. The differences between the roads are considerable; some are motorways carrying heavy traffic between major cities in the south, whereas others are relatively small rural roads in remote parts of northern Norway. While some of the projects have mobility as their main focus, others explicitly express the desire to foster tourism. These roads are all built primarily for purposes of transportation, however, and are therefore part of the everyday aesthetics of the Norwegian Public Roads Administration, rather than the more artistically ambitious tourist roads discussed by Kampevold Larsen in Chapter 13.

The individual guidelines vary considerably in extent, style, and content, and are composed by different groups of people with different disciplinary backgrounds. Furthermore, they concern different road construction projects, with widely differing ambitions, goals and financial resources available. While some ideas and considerations are universally present in the guidelines, others can be found in some, but are conspicuously overlooked or even refuted in others. Some of the guidelines seem to find considerable scope for aesthetic experience and aesthetic improvement in the road projects presented, and come across as highly ambitious, whereas others seem to be characterized by an air of slight resignation, and focus on only partially solved problems and trade-offs between different concerns. Nevertheless, there is a certain uniformity between the guidelines that justifies reading them as an expression of an aesthetic, although a somewhat flexible one.

THE AESTHETICS OF ROAD PROJECTS

The guidelines often begin by setting out their overall goals or philosophy, sometimes in the form of a separate chapter, sometimes in the introduction, and sometimes as a set of overarching principles, ideals or 'good practices' for design. The general aim is to construct a beautiful road, which should be experienced as a positive value for road users as well as outside observers. Many of the documents declare a double focus: On the one hand, the road should be seen as an integrated whole, with its own specific identity, on the other, the road is to present the local landscapes in their own right and to provide variation and diversion for road users. The identity of the road as a sphere in its own right is preserved through the use of recognizable materials and design principles – typical examples are homogenous central reservations, guardrails and roadsides.

The experience of variation is often described in terms of sequence and rhythm, or even with reference to film, dance and music; terms reminiscent of Appleyard, Lynch and Myer's classic description of road travel.[5] The challenges for road designers are also described in terms of striking the right balance between these aspects, so as to avoid on the one hand monotony, on the other anarchy, chaos and obtrusive specificity.[6] The experience of the road users must, according to most guidelines, be composed as a harmonious whole with recurring themes and climactic moments followed by calm intervals.

The guidelines aim to make the character of each landscape type stand out, and adapt to local conditions. For this purpose, a landscape analysis is usually carried out, and the surrounding terrain is divided into sequences based on landscape characteristics, such as forests, agricultural land, mountains, hilly or flat sequences, semi-urban areas, areas around a lake, etc. It is seen as desirable to reinforce the characteristics of each type of landscape, so as to awaken the travellers to the nature of the nature – and culture – through which they are passing. This can be ensured for instance by opening up a view of the sea where a road passes along a fjord, or, as in one of the guidelines, the alignment of the road should underline that the road is situated on a moraine.[7] More subtle measures include the use of local materials, or reinforcing the character of the landscape through the symbolism of instalments such as rest stops that are designed to resonate with the 'theme of the land' or traditional local architecture. In one of the more striking instances, a section of the road is defined as 'the Troll Forest', where a forest with spectacular boulders is to be illuminated at night.[8]

There are also other typical goals that are set out in most of the guidelines. One is a general commitment to preserving the integrity of the landscapes. Another is to create a simple, logical, legible road system with a clear, hierarchical structure, something which is also seen to be important from the perspective of road safety. Simplicity is also presented as an aesthetic ideal, however, so that the road should appear as sleek and uncluttered as possible, avoiding too many signposts, trying to limit the use of guardrails and downplay the size of road lightning. It is often emphasized that the road should appear 'neutral' and 'plain',[9] and that one should avoid measures that are 'unnecessary', 'alien' or 'domineering'.[10] The road should appear 'modern' or, in some cases 'timeless',[11] and should not require too much maintenance, but be based on materials and solutions that are 'durable'.[12]

LANDSCAPES AND ROAD ALIGNMENT

One of the important principles set out in most of the guidelines is that road alignment should harmonize with the surrounding landscape. Generally, the road should be situated in such a way as to minimize encroachment. The scale of the road should ideally be subordinate to the scale of landscape, and conform to its natural forms.[13] It should, as one of the documents put it, appear 'as though it has always been there'.[14] It should strike the traveller as well as the distant observer as a natural part of the landscape, not as an incursion or a violent intervention: 'The road should appear to belong to the landscape. The magnificent and untouched landscape should be at the centre of the traveller's experience'.[15] A few of the guidelines present the ambition that the road should appear to be 'soft-spoken'.[16] It should follow the contours in the landscape, and move along its natural forms. The ideal aesthetic form of the road is sinuous, with smooth curves, while the straight road is considered to

provide an unpleasant perspective, with the opening ahead creating an unsightly division of the landscape. It should not be situated in the middle of flat terrain, and it should run along natural divisions of the landscape. The linear aspect of the road should be downplayed, for instance through the use of vegetation, if alignment alone will not serve the purpose sufficiently well.

The road should not be experienced as a dominating feature, neither for the road users, nor for those who watch the landscape from a distance. The road should also open up the terrain through which it moves. It should make the different landscapes available to the road user in a direct manner, and create the impression of openness and access to the surroundings. Any barriers and barrier effects should be avoided, as well as constructions, such as short bridges, which may create the effect of narrowing the space of the road. The road user should not have the experience of being fenced in or closed off from the landscape.

If the road passes through steep terrains, for example, one may therefore consider placing the carriageways on different levels, so as to minimize the impression of encroachment, from the road as well as from the distance, and also open up the view to both sides. And in keeping with the ambition to avoid unnecessary clutter, roads should be placed and designed in such a way that guardrails can mostly be avoided.

MOUNTAINS, ROCKS AND TUNNELS

Rock cuttings are usually seen as a necessary evil in the guidelines: The exposure of the naked rock, especially if it is bland and monotonous, should be avoided as they appear as unsightly 'wounds' or scars in the landscapes.[17] If the cuttings are small, they can in some instances be covered completely by earth and vegetation. Rock cuttings which expose more interesting geological features may sometimes be turned into a consciously designed decorative element, and in some cases 'cuttings may also appear as precise sculptures'.[18] In order to

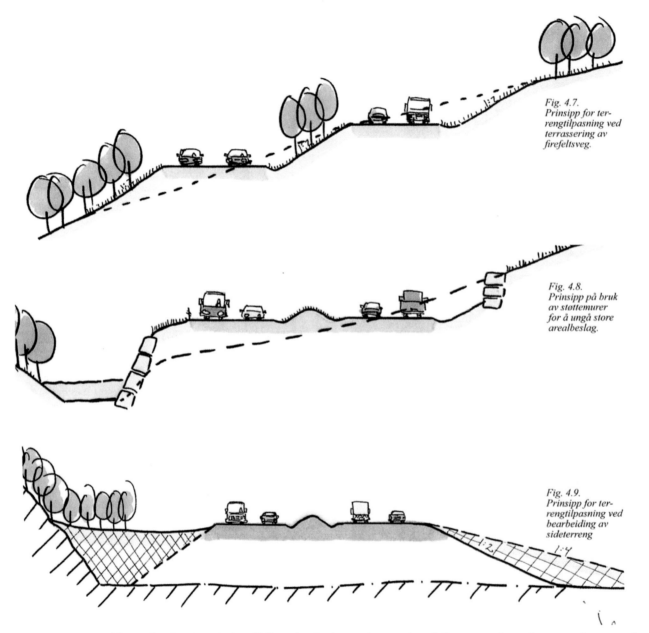

Fig. 4.7.
Prinsipp for ter-
rengtilpasning ved
terrassering av
firefeltsveg.

Fig. 4.8.
Prinsipp på bruk
av støttemurer
for å ungå store
arealbeslag.

Fig. 4.9.
Prinsipp for ter-
rengtilpasning ved
bearbeiding av
sideterreng

Figure 15.1 Norwegian Public Roads Administration's guidelines showing how an improved road alignment reduces the encroachment on the natural landscape

Formingsveileder Vestkorridoren E16 Hamang – Skaret (Oslo: Statens vegvesen, 2005). Reproduced with permission of the Norwegian Public Roads Administration.

Figure 15.2 Recommendations to avoid road cuts

The guideline reads: (Left image) 'View of the road from Skatvedtfeltet as the road was planned in the regulation plan. The high cutting on inside of the road is clearly exposed towards the residential area.' (Right image) 'View of the road from Skatvedtfeltet as it will now be. The road is hidden by a planted embankment which will also reduce noise.' *Veileder for estetisk utforming, Oslofjordforbindelsen* (Oslo: Statens vegvesen, 1997). Reproduced with permission of the Norwegian Public Roads Administration.

soften the impression of the more usual kinds of cuttings, it is frequently recommended that cuttings should be as low as possible, be rounded at the top, and be softened with land fills and vegetation, and broken up by niches of vegetation so as not to appear monotonous. In general, high cuttings are seen to have an oppressive effect, especially if there are cuttings on both sides of the road, which can create the feeling of driving between walls or in a ditch. The cuttings should therefore not be too steep, but be softened by fillings gently sloping towards the road. In general it is seen as desirable to reduce the visual impression of infringement of or violence to the landscape.

Tunnels provide a different set of challenges. They will inevitably stand out as encroachments in the terrain, but even so, they should fit harmoniously into the overall structure of the road. The cuttings necessary in front of the tunnel should be reduced to a minimum through the careful location of the opening, and fills and stonewalls, ideally using local stone, should cover the scars left by construction work.

TREES, BUSHES AND VEGETATION

While these guidelines do not, for the most part, favour the use of plants and flowers for purposes of decoration, or as part of a process of 'beautification',[19] vegetation still plays a number of important roles in the design of roads. They contribute to 'anchoring' the road in the landscape. For one thing, vegetation is essential for quickly masking the traces of recent development work. It is frequently advised that areas close to the road should be re-sown as quickly as possible. Another function of vegetation is to reinforce the sense of the local. This is achieved through the use of indigenous, ideally 'typical' plant species, partly through a process of re-vegetation, which involves preserving and reusing the top-soil, exploiting 'nature's capacity to repair itself after incursions'.[20] Indigenous plants may also be sown, if this is necessary. Even among indigenous plants, however, care must be taken to achieve the right kind of effect; as when one of the guidelines states that while '[c]ommon wormwood tends to give a shabby impression', '[h]eather and other low plants make for a beautiful and tidy embankment'.[21]

The local vegetation is not only a tool, it is also seen as a given; the raw material with which the landscape architects work and part of the landscape which is to be presented to the passing traveller. A pine forest, for instance, is a feature that provides variety but that may also appear dark, monotonous and uninviting, appearing as a wall along the road and thus being a problem as well as a resource. Thinning of the forests is suggested so as to increase the sense of light and openness, or perhaps to draw attention to interesting, sculptural trees or groups of trees.

One feature of vegetation that distinguishes it from the hard rock of the mountains and land formations, is its flexibility. It is possible to modify the appearance of the landscape quite profoundly by simple and low-cost measures such as thinning a dark forest, or cutting down trees to expose a view of the sea, a river, or an interesting building. Vegetation may also be used to actively construct a more varied and interesting landscape, as when one employs 'mass planting of indigenous species with the aim of constructing nature',[22] or sculptural single trees may be preserved or groups of trees are planted to serve as landmarks, milestones or diversions along the road.

As lines of trees are considered very dominant features, they should be applied with care, and follow the natural lines of the landscape. Colourful flowers in a meadow may be sown so as to present a striking view and a contrast to an otherwise bleak landscape.

Then again vegetation can be used to cover up in a more comprehensive way than just through re-growth on exposed surfaces, or it can take over the functions of more obtrusive kinds of materials: Trees may be planted to screen the view of nearby roads or unsightly industrial or commercial centres, in order to avoid visual and informational confusion. Some of the guidelines advocate the use of living noise walls – hedges of trees or bushes can serve the same functions as wooden or glass structures, and thus reduce the impression of disturbance to the landscape, and indeed to its inhabitants.

Vegetation can also improve the appearance of other installations, as when it is recommended that trees are planted behind road signs, to provide a screen against which the signs can be easily read, and reduce the appearance of clutter, or that climbing plants can be used to partly cover rock cuttings. In short, vegetation may 'open up to and frame views that are worth seeing for road users, and screen for what is experienced as 'visual noise'. Border areas and transitions where 'something is going on' are denoted with vegetation'.[23]

BRIDGES AND CROSSROADS

Crossing bridges are, as is frequently pointed out, very visible elements of the road, and should therefore be given 'timeless' aesthetical appearance, be properly aligned with the landscape, and avoid unnecessary infringement of the terrain. Bridge constructions are usually planned to give the impression of 'lightness', again avoiding the impression of oppressive weight and closure, and encourage the feeling of openness and flow. Supporting columns should therefore be slim and guardrails light – sometimes partly transparent. There is often an ambition to secure a degree of variation between bridges when it comes to design, sometimes the guiding principle is that the local environment should be decisive, but, at the same time, a certain measure of visual uniformity is seen as necessary in order to retain the road's identity and visual homogeneity.

Crossroads are exempt from the general requirement for naturalness. They are occasionally presented as diversions, or as points of the journey that have the potential to stand out as a *place* or serve as milestones. They can be given their own identity, or be turned into more park-like installations. The detailing of the decor is often intended to signal something about the place, as when a crossroads near a town can be given an urban character through the use of flagstones and tiles. The use of plants and trees can give the individual crossroads its own identity. One of the guidelines suggests giving each of the road's crossroads its own seasonal identity, through using trees and plants that are associated with different seasons, such as firs for winter, and lilac for spring. The crossroads should still adhere to the rules of simplicity and legibility, though, so the full complexity of the crossroads should not always be apparent to the road user, and nearby roads may be screened off by rows of trees, for instance.

THE EXPERIENCE OF THE ROAD USER

Several of the most recent guidelines include as a separate consideration 'the experience of the road user', and a reference to a study of the same name published by the Norwegian Public Roads Administration in 2005. The term refers to a conception of the experience of the road as an intersection between aesthetic enjoyment and road safety. The starting point for these considerations is that the well-designed road may also contribute to reducing monotony and thus sleepiness – which is an important risk factor in road traffic, and one that is not easily reduced through traditional measures such as enforcement. This is not a new idea. In 1969, Sylvia Crowe stated that 'Endless plugging along a featureless road, or along one which repeats a series of set features, is dull and therefore dangerous'.[24] This means that the good road should not simply appear to be aesthetically pleasing, but also stimulating. In the booklet 'The experience of the road user – Theory and methods' it is thus established that it is desirable to have one attraction – also referred to as one stimulus – every three minutes in order to keep the road user alert and interested.[25] Parallels are drawn to films and popular music, where similar intervals can be found. The road should have its own 'rhythm', where the driver is diverted, but not overwhelmed. These attractions, it is stated, can serve as milestones, and thus mark one's progress through the landscape, creating a sense of place. On the basis of these findings, a system of notation is developed, where stretches of road are divided according to whether they are deemed to be 'monotonous', 'in between', or are 'stimulating'. Maps of the roads are then produced, where the legend includes symbols for elements such as landmark, sculptural element, milestone, barrier, view, and chaos. A preliminary investigation is cited which found that stretches defined as 'stimulating' had a 23 per cent lower accident frequency.

What emerges is a concept of experience that conflates the interests of aesthetics and road safety, and proscribes a certain kind of landscape, or a landscape with a particular experiential rhythm. Based on this presumption, several of the guidelines set out to find where such experiences can be located along the road in question, whether they are already present, need to be 'cultivated' or 'strengthened' or must simply be introduced. The stimuli suggested range from traditional ones, such as views of landscapes or buildings, to more innovative measures such as illumination and use of art along the road.[26] Cultivation can take the form of removing vegetation to bring out a view, endowing crossroads with their own specific identities, or bringing out more clearly the characteristics of certain stretches of road. New stimuli, on the other hand, can be spectacular kinds of vegetation, constructions such as characteristic bridges, or art in the form of lightning, sculpture or land art, for instance. One of the guidelines includes a separate appendix on the possibility of turning the road into a showcase for art 'the world's longest art gallery' and an evaluation of the possible safety effects of this measure.[27] Art, in this context, is presented partly as a life-saving measure, partly as a possibility for advertising the road and the region, and even an arena for exhibiting contemporary Norwegian art.

THE INTERNATIONAL LANDSCAPE, THE NATIONAL LANDSCAPE AND THE LOCAL LANDSCAPE

Most of the ideals, principles, and measures propagated by the guidelines are far from being new – they do not differ markedly from those set out in the Public Road Administration's handbook *The road in the landscape* from 1978 – or unique to Norwegian road building practices.[28] Most of them can be found in earlier literature on roads and landscapes, such as works by Sylvia Crowe, Geoffrey Jellicoe, or Tunnard and Pushkarev,[29] and to practices in other countries, such as the German Autobahn and the American parkways.[30] The documents advocate roads that are subservient to the landscape and sinuous, smooth curves and embankments that link naturally to the surrounding terrain, vegetation that contributes to linking the road to the landscape, and light, slender installations. Further, one should avoid clutter, barriers, decoration, the excessive use of guardrails and noise walls, counteract the feeling of being locked in, and camouflage major interventions in landscape. Many of these principles can be traced back all the way to English picturesque theory by way of American landscape architecture, especially Frederick Law Olmsted.[31] There seems to be little divergence from these international and time-tested tenets of road design. As noted by Mauch and Zeller, 'technological and cultural transfer have contributed greatly to a convergence of engineering and aesthetics in different part of the world' where road construction is concerned.[32] So the question is; what kinds of individual landscapes emerge from these shared practices of constructing beautiful roads? How can these standardized and international principles for road construction serve to present the landscape as a specifically national, regional or local landscape?

The first answer that suggests itself is that one of the international practices adopted by the Norwegian road planners is aimed exactly at reinforcing the sense of the local or the national. Since the design of the road is based on a prior characterization of the landscapes through which the different sections of the road run, the landscapes are already interpreted, and when the road is designed, the importance of exhibiting the landscape will take the form of accentuating a meaning that is already inscribed into it. If the landscape is a 'fjord landscape', a view should be opened up to make this clear to the road users; if the road is situated on a moraine, this should be highlighted by the situation of the road; if the road runs close to a town or village, vegetation can be more park-like and constructions could be urban in character. One of the guidelines states that the quality of the local landscape must be 'strengthened and cultivated'.[33]

Another way in which the landscape emerges as local or national has to do with what views are presented to road users and what is hidden from view. In accordance with international principles for road design, certain views are deemed 'ugly' or 'chaotic' and thus wholly or partially shielded from the road. These are often commercial centres or industrial constructions. The views that are promoted, on the other hand, are typically views of landscapes, seas, and rivers (it is often mentioned that the view of water is a source of visual enjoyment – or simply that 'water is an important element and attraction'[34]), or of historical or architecturally important buildings. The views again reinforce an understanding of the landscape, nation and region that is already current.

In almost all of the documents, the importance of using vernacular (or at the very least national) plants is also stressed, although the reasons provided for this are not always merely aesthetic; in some cases, reference is made to avoiding the migration of alien plants which can be a consequence of road construction. The practice is thus also linked to environmental concerns. However, it is usually suggested that a similar practice is adopted when it comes to the building of stone walls, for instance, which should be constructed in local stone.

Furthermore, when it comes to constructions such as rails, bridges and tunnels, there is a strong emphasis on stone, wood and steel that is recognizable from contemporary Norwegian architecture. This aspect is particularly striking in the design of rest stops, which are often explicitly designed to reflect the local landscape and culture. In one of the plans, different thematic schemes are developed for different stretches of the road, so that the wooden furniture chimes in with the surrounding forest, while a 'maritime' design is introduced where the landscape opens up to the fjord.[35] In another document, rest stops are to be designed in as modern interpretation of the local architectural tradition of small wooden houses, which is strongly associated with the idyllic character of the local landscape, which is one of the most popular seaside areas of Norway.[36] Rest stops are also often situated where a view opens up to a typical landscape, and sometimes vegetation is removed and tracks are constructed to provide direct access to the countryside. The sum of these design practices could be seen as a part of the process that Sörlin terms 'articulation of the territory'.[37]

CONCLUSIONS

One of the main objectives of the road planning apparent in the guidelines is to avoid two related effects; on the one hand, that the road appears to be separated from the landscape, on the other; to prevent the sensation of driving in a corridor or tunnel. The transition between road and surrounding should always be seamless, and enable road users to experience themselves as part of the terrain. However, modern requirements for safety and mobility seem to favour the construction of the road as an enclosed space. For one thing, the roads are typically planned for speeds between 80 and 110 km/h. Modern principles of road safety in general, and the Vision Zero for road safety adopted in Norway especially, suggest that at such speeds, different groups of road users should be separated.[38] Another tenet of recent road safety thinking is that road sides should be *forgiving*, meaning the terrain closest to the road should be a 'safety zone', where a car will not hit hard objects, such as rocks or trees. Where such a zone cannot be established, the guidelines reluctantly allow for using safety rails, which, again, form a solid barrier between the road and its surroundings.

Further, major roads will often require measures to counteract noise, usually in the form of embankments (typically the preferred solution), or, where space does not allow for this, noise walls. These walls are among the most problematic aspects of road design, as they do not merely create a psychological barrier, but also a very visible one, which makes for monotonous views. The ambition is usually to render the walls as unobtrusive as possible, they are often to be withdrawn from the road so as to allow for vegetation which at least partially covers them, and painted in dark or neutral colours.

Frequently, it is the task of the landscape architect to disguise the fact that the road is a separate sphere and conceal all the measures taken to isolate the road from its surroundings. It is in a sense a re-integration of the road in the landscape, which is more demanding the higher the speed. The ideal is that the road and landscape should be experienced as 'natural', yet they are both highly constructed and planned in great detail, in some cases, it might even be justified to think of this in terms of a 'naturalness effect'.

The road should be seen as part of landscape, yet one is constantly being propelled forward, and one's progress is carefully measured by landmarks and attractions. It is opening up the landscape, but constantly leading the spectator's gaze, and masking the extremely strict limitations and restrictions that the road users are actually subject to. The landscape of speed and the landscape of still or slowness, that are typically associated with place rather than space, are superimposed on top of each other. The landscape of speed is necessarily derivate; places and immobile objects and views are what create the sense of speed, serving as – as it is frequently stressed in the guidelines – milestones and landmarks.

The goal of landscape architects, therefore, is often to create or reinforce a sense of place in order to construct a sense of movement. This way of organizing roads seem to give characteristic and recognisable places an unfamiliar air of instrumentality; places should be created in order to create a sense of progress, and keep the driver mildly diverted. Thus the debate over whether the motorway should be understood as a non-place seems to miss the mark in this context; for a 'good' (aesthetically as well as ethically) experience of travel to come about, places are necessary condition, although not experienced (by the road user) through traditional criteria such as limits or habitability.[39]

ENDNOTES

1   See for instance Christof Mauch and Thomas Zeller (eds), *The World beyond the Windshield. Roads and Landscapes in the United States and Europe* (Athens: Ohio University Press, 2008); Peter Merriman, *Driving Spaces* (Oxford: Blackwell, 2007); and 'A new look at the English landscape': landscape architecture, movement and the aesthetics of motorways in early postwar Britain'. *Cultural Geographies* 13/1 (2006), pp. 78–105; Mary E. Myers, 'The Line of Grace: Principles of Road Aesthetics in the Design of the Blue Ridge Parkway'. *Landscape Journal* 23/2 (2004), pp. 121–140.

2   Prior to this date, landscape architects were occasionally commissioned to work for the Norwegian Public Roads Administration, so this should not be taken to mean that aesthetic considerations were entirely absent from Norwegian road planning before this date. Engineers also adhered to many of the same ideals as discussed in this article. As an indication of this, 'Synspunkter på vegestetikk' (Notes on road aesthetics), an article from *Norsk Vegtidsskrift* [Norwegian Road Journal] 37/10 (1961), pp. 173–176, written by senior engineer Svein Nesje explains how the road should be adapted to the landscape, and how the sinuous curve is the most beautiful shape for a road.

3   There exists no central archive of such guidelines, so the selection discussed in this chapter – which is unlikely to be absolutely exhaustive – is collected with the assistance of landscape architects in the different regions of the Norwegian Public Roads Administration. I am grateful to them all for having made the material available to me, and especially to Sunniva Schjetne in the Road Directorate. I would also like to express my gratitude to Ingerlise Amundsen, who has provided valuable input on Norwegian road planning and on relevant literature.

4   It should be noted that these guidelines are planning documents, and the finished results may sometimes diverge considerably from the goals stated. This is an attempted analysis of a discourse and practice of *planning* the aesthetically pleasing road, rather than the actual material practice of building it. The guidelines, furthermore, are not free-standing, but part of a complex hierarchy of planning documents published prior to major road constructions. They also relate to other, more fundamental documents that place quite severe restrictions on their agency. Central among these documents are the National Plan for Transportation; the Norwegian Public Roads Administration's norms for road construction; and national and regional environmental regulations.

5   Donald Appleyard, Kevin Lynch and John R. Myer, *The View from the Road* (Cambridge, Massachusetts: MIT Press, 1964).

6   *Veileder for estetisk utforming. Oslofjordforbindelsen* (Oslo: Statens vegvesen, 1997).

7   *Visuell veileder for 4-felts E6 gjennom Østfold. Utforming av veg- og sidearealer* (Oslo: Statens vegvesen, 2002).

8   *E6 Gardermoen – Biri Formingsveileder* (Oslo: Statens vegvesen, 2006, revised 2009).

9   *Formingsveileder E18 Bommestad-Sky* (Oslo: Statens vegvesen, 2009), *Formingsveileder E6 Trondheim-Stjørdal, Parsell Trondheim fjelltunnel og dagsone øst* (Oslo: Statens vegvesen, 2008), *Formingsveileder Vestkorridoren E16 Hamang – Skaret* (Oslo: Statens vegvesen, 2005), *E6 Gardermoen – Biri Formingsveileder* (Oslo: Statens vegvesen, 2006, revised 2009), *Veileder for estetisk utforming. Oslofjordforbindelsen* (Oslo: Statens vegvesen, 1997).

10 *Formingsveileder Vestkorridoren E16 Hamang – Skaret* (Oslo: Statens vegvesen, 2005), *Estetisk veileder. Ny*

E18. 4-felts motorvei Grimstad-Kristiansand (Oslo: Statens vegvesen, 2003, revised 2005), *Formingsveileder E 18 Bommestad-Sky* (Oslo: Statens vegvesen, 2009), *Formingsveileder E6 Øst, Trondheim- Stjørdal. Parsell: E6 Værnes-Kvithammar* (Oslo: Statens vegvesen, 2008), *Formingsveileder E6 Trondheim-Stjørdal, Parsell Trondheim fjelltunnel og dagsone øst* (Oslo: Statens vegvesen, 2008).

11 *E6 Biri-Otta. 'Vi byr på Gudbrandsdalen. Formingsveileder Lillehammer nord – Otta* (Oslo: Statens vegvesen, 2008).

12 *Formingsveileder E18 Gulli-Langåker* (Oslo: Statens vegvesen, 2009), *E6 Biri-Otta. 'Vi byr på Gudbrandsdalen'. Formingsveileder Lillehammer nord – Otta* (Oslo: Statens vegvesen, 2008).

13 Amundsen, Selberg and Lundebrekke point out that in practice, however, this has become increasingly difficult due to the requirements specified in the national norms for road construction. Ingerlise Amundsen, Knut Selberg, and Egil Lundebrekke, *Veg og landskap* (Trondheim: Universitetet i Trondheim, Norges Tekniske Høgskole, Institutt for veg- og jernbanebygging, 1995).

14 *Rv 23 Oslofjordforbindelsen – en snarvei under Oslofjorden* (Oslo: Statens vegvesen).

15 *E10 Raftsundet Øst – Ingelsfjordtunnelen. Håndtering av landskapet i plan- og byggeperioden* (Oslo: Statens Vegvesen, 2004).

16 Ibid., *Formingsveileder Vestkorridoren, E 16 Hamang- Skaret* (Oslo: Statens vegvesen, 2005), *Riksveg 17 Tverlandshalvøya: Håndtering av landskap i plan- og byggeperioden* (Oslo: Statens Vegvesen, 2006), *Veileder for estetisk utforming. Oslofjordforbindelsen* (Oslo: Statens vegvesen, 1997).

17 *Rv 23 Oslofjordforbindelsen – en snarveg under Oslofjorden* (Oslo: Statens vegvesen).

18 *Visuell veileder for 4-felts E6 gjennom Østfold. Utforming av veg- og sidearealer* (Oslo: Statens vegvesen, 2002).

19 On the debate around 'beautification' of roads in the UK, see Peter Merriman, '"Beautified is a vile phrase". The politics and aesthetics of landscaping roads in pre- and postwar Britain', in Mauch and Zeller (eds), *The World beyond the Windshield*.

20 *E10 Raftsundet Øst – Ingelsfjordtunnelen. Håndtering av landskapet i plan- og byggeperioden* (Oslo: Statens vegvesen, 2004).

21 *Visuell veileder for 4-felts E6 gjennom Østfold. Utforming av veg- og sidearealer* (Oslo: Statens vegvesen, 2002).

22 *Formingsveileder for vegene i Namdalsprosjektet* (Oslo: Statens vegvesen, 2002).

23 *Formingsveileder, E6 Øst, Trondheim – Stjørdal, Parsell: E6 Værnes – Kvithammar* (Oslo: Statens vegvesen, 2008).

24 Sylvia Crowe, *The Landscape of Roads* (London: The Architectural Press, 1960).

25 *Trafikantens opplevelse. Teori og metode* (Oslo: Statens vegvesen, 2005).

26 *E6 Biri-Otta 'Vi byr på Gudbrandsdalen' Formingsveileder Lillehammer Nord-Otta* (Oslo: Statens vegvesen, 2008), *Vegpakke Helgeland – visuell veileder (Statens vegvesen, 2006), Formingsveileder E18 Bommestad – Sky* (Oslo: Statens vegvesen, 2009), *Formingsveileder E18 Gulli – Langåker* (Oslo: Statens vegvesen , 2009), *E6 Gardermoen – Biri Formingsveileder* (Statens vegvesen 2006, revised 2009).

27 *Visuell veileder for 4-felts E6 gjennom Østfold. Utforming av veg- og sidearealer* (Oslo: Statens vegvesen, 2002).

28 *Vegen i landskapet* (Oslo: Statens vegvesen, 1978).

29 Christopher Tunnard and Boris Pushkarev, *Man-Made America: Chaos or Control?* (New Jersey: Yale University Press, 1963).

30 Zeller, *Driving*, and 'Building and Rebuilding the Landscape of the Autobahn, 1930-1970', in Mauch and Zeller (eds), *The World beyond the Windshield*.

31 Mary E. Myers, 'The Line of Grace: Principles and Road Aesthetics in the Design of the Blue Ridge Parkway', *Landscape Journal* 23/2 (2004), pp. 121–140.

32 Christof Mauch and Thomas Zeller, 'Introduction', in Mauch and Zeller (eds), *The World beyond the Windshield*.

33 *Visuell veileder for 4-felt E6 gjennom Østfold* (Oslo: Statens vegvesen, 2002).

34 *Estetisk veileder. Ny E 18 4-felts motorveg Grimstad – Kristiansand* (Oslo: Statens vegevesen, 2003 revised 2005).

35 *Veileder for estetisk utforming. Oslofjordforbindelsen* (Oslo: Statens vegvesen, 1997).

36 *Estetisk veileder. Ny E 18 4-felts motorveg Grimstad – Kristiansand* (Oslo: Statens vegevesen, 2003, rev. 2005).

37 Sverker Sörlin, 'The articulation of territory: landscape and the constitution of regional and national identity', *Norsk Geografisk Tidsskrift* 53 (1999), pp. 103–111.

38 See Beate Elvebakk, 'Vision Zero: reshaping road safety'. *Mobilities* 3/2 (2007), pp. 425–441.

39 Edward S. Casey, *The Fate of Place. A Philosophical History* (Los Angeles, California: University of California Press, 1997).

16 Enfolding and Gathering the Landscape: A Geography of England's M1 Corridor

Peter Merriman

[...] the State needs to subordinate hydraulic force to conduits, pipes, embankments, which prevent turbulence, which constrain movement to go from one point to another, and space itself to be striated and measured, which makes the fluid depend on the solid, and flows proceed by parallel, laminar layers. The hydraulic model of nomad science and the war machine, on the other hand, consists in being distributed by turbulence across a smooth space, in producing a movement that holds space and simultaneously affects all of its points, instead of being held by space in a local movement from one specified point to another.[1]

[...] even though the nomadic trajectory may follow trails or customary routes, it does not fulfill the function of the sedentary road, which is to parcel out a closed space to people, assigning each person a share and regulating the communication between shares.[2]

In outlining their nomadology in *A Thousand Plateaus*, Gilles Deleuze and Félix Guattari famously explored the interconnections, overlappings and movements associated with minor, nomad science *and* state science, where the former is characterized by a hydraulic or vortical model of openness, fluidity, smoothness, heterogeneity and becoming, and the latter by striation, sedentarism, ordering, regulation and fixity. While the modern Western road might appear to be characterized by such fixity, regulation and channelling – closely bounded by privately-owned, enclosed lands, and monumentally fixed in concrete, steel and tarmac – not all roads are or have been this way. Anthropologists have shown how non-Western peoples often have very different conceptualizations of what roads are and how and where they are demarcated, as well as how roads and vehicles should be used, particularly in desert, mountain, tundra or icy areas, where roads may collapse, be covered over with sand or snow, swept away by water, or liable to melt.[3] Modern surveying and road construction techniques, coupled with Western models of private property ownership, have led to a much clearer demarcation of the edges and spaces of roads, but in this chapter I want to suggest that roads – and particularly motorways – have a more complex and diffuse geography than it may at first appear.

In this chapter I focus on England's M1 motorway running between London, the Midlands, and Yorkshire.[4] At first sight the motorway may appear to have a simple linear geography, forming a transport corridor between London and Leeds, but this linearity is punctuated and disturbed by the distinctly nodal geography of motorway junctions, which are positioned every 4 to 10 miles.[5] In this chapter I examine these complex corridoring effects, focusing on the topological and spatial patterns associated with the linear, nodal, material, imaginative and discursive geographies of motorways. Of course, urban planners, transport geographers, economists, logistics scholars and regional scientists have written fairly extensively about transport corridors and their effects on regional development, economic investment and related matters, but there have been almost no in-depth studies examining their cultural and social dimensions.[6] In this chapter I very loosely draw upon the concept of topology as a metaphor for understanding the non-linear, non-Euclidean and relational aspects of motorway corridors. Motorway corridors may be characterized as turbulent spaces which are in process and are constantly formed and emerge through the flows of bodies, vehicles and other materials across their borders. I approach motorways as 'places' rather than as archetypal 'non-places',[7] where places are conceptualized as open, dynamic, relational, processual 'mobile effects': 'a non-representation that is mobilized through the placing of things in complex relation [sic] to one another and the agency/power effects that are performed by those arrangements'.[8] Places, like landscapes, should be seen to be continually ordered, practiced and placed through the folding together of different materials, atmospheres, spaces and times into a complex topology or, as Marcus Doel has put it, a 'scrumpled geography'.[9]

LOCATING THE M1

In what sense was the M1 seen to be part of the surrounding landscape, the countryside through which it passed? Clearly local residents were well aware of its location and construction. Some protested against its routing, while others complained about the disturbance caused during construction.[10] Landscape architects argued that the road must be tied into the surrounding landscape through careful design and planting.[11] But for many motorists, the new motorway was a somewhat dislocated space – echoing arguments made by anthropologist Marc Augé and architect Paul Andreu.[12] Many motorists did not know how to find this

Figure 16.1 The cover of Motorway, *a pull-out guide to the M1 motorway published in* The Motor magazine, *4 November 1959*
Reproduced from *The Motor* magazine (now *Autocar* magazine) by permission of Haymarket Publishing.

new road, which was absent from older road maps. Others relied on the special guides which were issued by motoring organizations, petrol companies, and the press to find the motorway, such as the Automobile Association's 'Guide to the motorway', which contained a map of the route, extracts from the Motorway Code, and details of how to get to the motorway.[13] On 4 November 1959 *Motor* magazine inserted a special pull-out guide into its magazine, which featured an annotated map of the route and junctions, and details of how to access the motorway 'from the South' (Figure 16.1).[14] All of these different guides 'placed' the motorway in relation to local roads and towns, guiding drivers to these spaces, and familiarizing them with the new styles of driving which were required. Other motorists traced the line of the motorway onto their own older maps. For example, on a late 1950s Shell Great Britain map I found in my old office at the University of Reading, the owner had hand-drawn the line of the M1 onto the map – improving its accuracy, extending its useful life, saving themselves money. Children, too, were encouraged to chart this new space, to map its geographies, to locate the motorway in the landscape. In a 1959 BBC Home Service for Schools radio programme, the narrator, well-known Welsh preservationist Wynford Vaughan-Thomas, encouraged listeners to map the spaces of the M1:

Now take a look at your map, if you haven't already done so, and see exactly where the motorway is. [...] it runs right up through the heart of England. [...] you can roughly trace its route by drawing a line from London to Birmingham which passes to the west of all these towns.[15]

The child/listener is encouraged to act as a pioneering geographer-cartographer, charting a new space, a new road, on their own map, as well as placing the motorway in the English landscape.

So, map manufacturers, motoring organizations and other cultural commentators attempted to locate the new motorway in relation to familiar roads and destinations, but the opening of a modern national motorway also encouraged observers to rethink the geographies of places *in relation to* the geographies of this new motorway. The M1, then, starts to reorient how commentators view rural Bedfordshire,

Buckinghamshire and Northamptonshire, as local sites and landscapes become gathered into and around the motorway and its junctions, being identified as places to be seen or visited from the road. In 1960 the RAC's *Guide and Handbook* included a list of 'where to stay and where to eat in the vicinity of the London-Birmingham Motorway'.[16] In this and subsequent guides we see the geographies of the surrounding landscape, particularly the geographies of local hotels, restaurants and tourist attractions, become gathered around the motorway, enfolded into its spaces, incorporated into its relational and topological geographies, located in relation to its junctions. The motorway becomes a way of organizing or relating features of the landscape, providing the kinds of corridoring effects or geographies discussed by Peter Bishop in his study of the Alice Springs to Darwin railway in Australia.[17] This 'organizing' or 'gathering' was particularly evident in successive editions of Raymond Postgate's *The Good Food Guide*, which was seen to be rapidly joining 'the AA or RAC handbook and maps' as 'part of the usual travelling library of numerous middle-class motorists'.[18] As early as 1961 Postgate started to mention the propinquity of the M1 in reviews of restaurants in villages such as Newport Pagnell, Dunchurch, and Kislingbury:

The Swan Revived, Newport Pagnell [...] It has a genuine interest in food and it is well worth turning off the racket and rush of M.1 to dine here or sleep in a Victorian bedroom.[19]

Dunchurch, Warwickshire – Dun Cow Hotel [...] Usefully placed near the M1, this is a fourteenth-century coaching inn which offers a warm welcome and good food. The glossy menu is not remarkable for its diversity [...].[20]

Villages, restaurants, inns, hotels and tourist sites/sights all become worth visiting due to their proximity to the M1. They are gathered around the M1, linked to this national transport corridor. The Duke of Bedford's Woburn Abbey is a case in point. In 1958 it was announced that the M1 would cut across 100 acres of the Duke's land at Ridgmont, but when asked if he was concerned, the Duke pointed out that the motorway would help bring important revenue to his fledgling tourist venture:

It will mean that Woburn will be within an hour's run of London and that will mean more visitors. [...] Besides, the Government are paying me well enough for the land.[21]

The Duke of Bedford had opened Woburn Abbey to the public in 1955 to help pay death duties, and in guides to the Abbey then and since, maps show quite clearly the route between the park and the M1 – enfolding the M1 into the geographies of the park, just as today's tourist signs bring the Abbey into the spaces of the motorway.

The view of the landscape from the motorway also starts to be identified as important. In 1968 Shire Publications produced a new kind of guide book for the nation's motorists. Earlier guides had tended to focus on single counties or regions, but this new guide book, Margaret Baker's *Discovering M1*, used the linear transport corridor of the M1 as its organizing or gathering principle (Figure 16.2).[22] *Discovering M1* was written as 'a glove-compartment guide to the motorway and the places of interest that can be seen from it',[23] and it repeated the style of earlier guides and itineraries published for stage coach and railway travellers. Descriptions of landmarks and topographic features – such as the aerials at Daventry and Rugby, and ridge and furrow fields near Crick – were printed opposite maps, photographs, and sketches of the route. All of these were designed to enliven and illustrate the journey and the landscape for passengers:

It is written for passengers – perhaps bored by the apparent monotony of a road devoid of strip development and place-name signs – and is arranged for easy assimilation at around 60 mph.[24]

Drawing upon the arguments of Marc Augé, one could suggest that the guide was designed to animate, enliven or translate a monotonous non-place; rendering the landscape legible for passengers.[25] But I have problems with such a simplistic reading, and instead prefer to think of the guide as a tool which helped to perform this place, enabling travellers

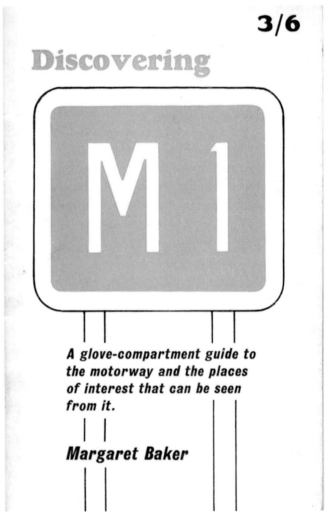

Figure 16.2 Margaret Baker, Discovering M1, *front cover*
Shire Publications, 1968. Reproduced by permission of Shire Publications, www.shirebooks.com.

to encounter and consume the landscapes of the motorway in different kinds of ways. This small-format guide was designed to reside in a glove compartment, and be taken out and read at a particular speed. At 60 mph, landscape features should appear and disappear as the passenger reads, glances and returns to the text.

M1 MODERN

The motorway was widely seen to be a modern force in the English countryside, bringing an international-metropolitan modernism into the English landscape and the imaginations of the British public. Cultural commentators celebrated the modernity of the new road and the experiences it fostered. As Ferrari's Formula 1 racing driver Tony Brooks remarked in *The Observer* in November 1959:

> To drive up M1 is to feel as if the England of one's childhood, the England of the British Travel and Holidays Association advertisements, is no more. This broad six-lane throughway, divorced from the countryside, divorced from towns and villages, kills the image of a tight little island full of hamlets and lanes and pubs. More than anything – more than Espresso bars, jeans, rock 'n' roll, the smell of French cigarettes on the underground, white lipstick – it is of the twentieth century. For all that, it is very welcome.[26]

The new motorway and its consumers are seen to bring a metropolitan chic into the landscape, extending the tentacles of the city. The motorway and surrounding landscape appear separated,[27] but in Brooks' view the motorway draws in, gathers and enfolds the metropolitan geographies and modern practices which are associated with London, at the southern end of the motorway.

This line of argument appears particularly strong in relation to the new service areas which were opened at Newport Pagnell and Watford Gap in August and September

1960. As one regular visitor reflected somewhat nostalgically in 1985, Newport Pagnell service area quickly became a 'place of pilgrimage for teenagers hoping for instant glamour':

> For young people, the new road was a concrete escape to a new kind of excitement. Along it, on a Saturday night, would swarm the Morris Minors, XK Jaguars and Norton motor bikes, eating up the miles at incredible speeds in search of the bright lights. Their destination? Mr. Forte's snack-bar on the M1... this cosy man-made island called out to Britain's youth, the generation of teenagers who did not know there was anything special about being young but forsook the coffee bars of Soho to spend Saturday night "doing a ton" on this long straight road.[28]

Charles Forte had become known for the large number of small, modern, 'popular' cafes, milk bars and restaurants which he operated in central London, but during the late 1950s and 1960s he branched out across the country with motorway service areas, motels, up-market restaurants and hotels. Newport Pagnell was a modern service area, an island in the countryside that was purposefully designed to be 'bright, modern and comfortable, to blend with the exciting conception of the motorway itself'.[29] For those not lucky enough to visit the motorway and service areas, postcards brought its spectacle into their homes (Figure 16.3).[30]

In many senses the M1 could be seen to have captured the popular imagination. A large number of motorists queued up at the junctions to try out their vehicles on Monday 2 November 1959. The motorway became a popular destination for Sunday afternoon family drives, while many others (particularly locals) stood on bridges to watch the traffic pass by. London Transport even ran special Sunday afternoon motorway bus trips throughout November 1959.[31] For a wide variety of observers, from children's writers and pop bands, to toy manufacturers and postcard publishers, the M1 emerged as an important cultural reference point, an exciting, modern spectacle which could help sell products

FORTE'S RESTAURANT

THE M.I. MOTORWAY.
Fifty Five Miles in Length,
it is spanned by 134 bridges.

ON THE M.1.

BLUE STAR

APPROACH TO THE SERVICE AREA

PEDESTRIAN BRIDGE OVER THE M.1.

Figure 16.3 'The M1 motorway', black and white postcard, ca. 1960–3
Private collection.

– folding the modernity of the motorway into people's everyday lives. In 1960 the London-based instrumental group the Ted Taylor Four released their new 7 inch single 'M1'. 'M1' is a rather unusual high-tempo instrumental track, resonating with the sound of the clavioline (an early electric keyboard).[32] The song was originally to have been titled 'Left hand drive' to appeal to the audiences the band performed to at American air bases in Britain, but the widespread publicity associated with the opening of the motorway persuaded the band to change the title. As two members recalled in 1992: 'The building of Britain's first "super road" the M1 and all its publicity made the change of name a sound commercial idea'.[33]

Two toy companies, American firm Marx and British firm Tri-Ang, featured artistic impressions of the M1 and Sir Owen Williams's distinctive bridges on the covers to their independently-produced, British figure-of-eight race-track sets, associating the excitement and modernity of this *new* motorway with their fairly standard toys (Figure 16.4). The motorway enabled both Marx and Tri-Ang to re-brand their sets, bringing the aesthetics, movement, excitement, freedom and speed of the new motorway to the young child, who could race *their* cars on *their* own motorway, whenever they desired.

Children's writers also recognized the excitement of the new motorway, and in two books published in 1961, the M1

Figure 16.4 Cover to Marx's 'Motorway' clock-work toy car set
Private collection.

appears as a space for a plot to unfold, a space of excitement which could be depicted for the reader. As with a wide variety of other representations and commentaries on motorway driving these children's books construct the motorway driver as a male and heroically masculine figure. The first of the two books, *The Mystery of the Motorway*, was written by Robert Martin, author of at least twelve other 'mystery' titles.[34] The M1 becomes the location of a mysterious car crash, in which a high-powered Mercedes overturns at over 100 mph. The driver dies, but a passenger escapes. The crash is witnessed by Mike Dance, a young engineer in his late teens/early twenties who is driving home from Birmingham to London. Mike, his young brother Jim, and a friend become amateur investigators in an attempt to solve the mystery of the motorway crash. Throughout the book the M1 serves as a space of excitement, modernity and danger, as well as of

mystery. The book acts as a social manual for boys entering their teens, drawing positive associations between motorway driving, speed, youth, and masculinity.

The second story was Bruce Carter's *The Motorway Chase*, published in Hamish Hamilton's Speed series.[35] The story's hero, Clive, is driving down the M1, 'dream[ing] of the day when he would be World Champion driver'.[36] Looking ahead, he sees that the police are stopping vehicles searching for an escaped convict. Suddenly, the fugitive breaks through the check-point in a stolen Ferrari and Clive gives chase in a racing Mini. The Ferrari turns off the motorway, and before long Clive catches up, overtaking the Ferrari and forcing it off the road; allowing the police to arrest the convict. Clive's skill as a driver is rewarded with the opportunity to try out for a racing team, and his dream of becoming world champion appears within reach, providing a link between everyday life and champion status for the ordinary boy at home, reinforcing and reconstructing his masculinity and desires:

> *After that chase, Clive knew he was a good driver. One day Clive would be World Champion driver.*[37]

The motorway is constructed as a democratic or meritocratic space of masculinity, a space 'where all men [...] can demonstrate their superior masculinity and self-worth'.[38] Every boy and man is presented with the possibility of moving from the speed limit-less M1 to the race track, and from the status of amateur to professional race driver. The spaces of the motorway become worked into narratives of British modernity, youth and masculinity. The motorway emerges as a space of speed, crime, excitement, fantasy, a space of masculine desire. *The Mystery of the Motorway* and *The Motorway Chase* enfolded the spaces of the motorway into the imaginations, subjectivities and reading-spaces of young children – particularly boys – extending the consumption of the motorway's modernity out into the homes of the public.

MOTORWAY REGIONALISM

The motorway was seen to be a modern force in other ways, and during the 1960s commentators reflected on how the M1 *might* transform – or *was* shaping – the urban, rural, local, regional and national geographies of the London to Birmingham corridor. Writing in *Country Life* in October 1966, Brian Dunning remarked on how little attention had been paid to 'the growing tendency for motorways to act as local boundaries – artificial rivers, so to speak', changing 'the basic shape of the countryside'.[39] Dunning observed how, despite the presence of farm access bridges and tunnels, land owners were exchanging land, such that 'the number of farmers cultivating land on both the eastern and western margins is falling year by year'.[40] The M1 and other motorways appeared to be reshaping England's agricultural and hunting landscape. Hunt masters became 'unwilling to take their packs [of hounds] near to the heavy traffic', and foxes were observed to 'set up home alongside the motorways where they are assured, not only safety, but of a plentiful supply of dead carrion'.[41] In August 1959, this bisection, this severance, led the Oakley hunt to give up lands to the Whaddon Chase hunt,[42] while the Pytchley hunt and Grafton hunt discussed the possibility of redrawing their boundaries so that they could hunt to the West and East of the M1, respectively.[43] For these huntsmen and farmers the motorway did not exert a 'tunnelling effect', rather it served as an impenetrable barrier to their activities.

Commentators also predicted that the M1 might reinforce the existing regional geographies of England and Britain. Economists and geographers argued that the first sections of the M1 would reinforce the industrial axis between London, the Midlands and the North-West, creating 'corridor' and 'tunnel' effects and drawing development away from more peripheral areas. The motorway would reinforce regional inequality, but it was also argued that it could reconfigure geographies within regions. Cultural and economic commentators suggested that the first sections of the motorway would 'gather in' or 'open up' areas of rural Bedfordshire, Buckinghamshire and Northamptonshire. Writing in October 1959, conservationist Malcolm MacEwen suggested that the M1 might reinforce the regional inequality of the South-East and lead to the urbanization of this transport corridor:

> *In Northampton [...] the Chamber of Commerce has suddenly awoken to the fact that the new London-Yorkshire motorway has created wonderful new opportunities for industry and commerce because Northampton lies only a mile or two to the east of the motorway, and about midway between Birmingham and London. Will the new road not also increase London's magnetism and the drift of population to the south-east? Could the new road not even*

become the central urban motor road of a new London-Birmingham conurbation?[44]

In *The South East Study 1961–1981*, the Ministry of Housing and Local Government presented the M1 as an important catalyst to urban expansion. Northampton was identified as an ideal candidate for major population expansion, while Bletchley – located adjacent to the M1 and on the main West Coast railway line – was styled as the ideal location for a new city.[45] Indeed, the Milton Keynes Development Corporation was subsequently established in May 1967, and in their city plan, published in March 1970, the M1 appears as one of two major motorways that would bind the city into the surrounding region (the other being a new urban motorway along the route of the A5).[46] The construction of the M1 sparked the imaginations of city planners and property developers, and in the past few decades economists and real estate developers have paid increasing attention to the kinds of economic activity and urban development generated by the motorway. In the pages of *Estates Times* and *Chartered Surveyor Weekly*, property developers and surveyors regularly report on the changing fortunes of particular towns, office developments and industrial sites along the 'M1 corridor'.[47] The 'M1 corridor' is styled as a linear and nodal landscape of actual and potential development sites and more-or-less successful towns. In such accounts the corridor may be seen to be practised through the movements and actions of multi-national companies, property developers, investors, workers and their services and goods, as well as motorists, and the corridor has been seen to act as both a barometer and shaper of fortunes in the region. Developers were particularly attracted to junction 18 at Crick, at the northern end of the first section of the motorway, where the M1, A5 and projected M6 came in close proximity. As Brian Turton explained in *The Geographical Magazine* in April 1978, the junction was styled as 'Britain's first motorway-orientated growth point', emerging as the location for several large depots and unsuccessful proposals to construct a £35 million 'privately

developed new town'.[48] In the 1990s, developers constructed DIRFT (the Daventry International Rail Freight Terminal) to the west of junction 18, at which Tesco, Royal Mail, Eddie Stobart, Wincanton, Exel and other large companies have all established distribution and warehouse centres, and which is actively promoted by its owners as a key site in England's important '"golden triangle" favoured for national distribution centres'.[49]

Later sections of the M1 were predicted to have similar economic and development effects, and in the June 1960 issue of *The East Midland Geographer*, University of Nottingham geographer Richard Osborne examined the likely impact of the projected second, Rugby to Doncaster, section of the M1 on the economy, industry and population distribution of the East Midlands.[50] Osborne predicted that the motorway would strengthen the North-South axis of the East Midlands region: speeding up cross-regional journeys, opening up new areas, fostering increased demand for housing, generating new commuter routes, influencing the location of industry. The M1 would reshape the geographies of this newly conceived region, and it was felt that East Midlands planners would have to control the motorway's effects on the region's villages, towns, cities and countryside.

Beyond these regional effects, the M1 has opened up the English landscape in other kinds of ways. Small villages and towns – including Toddington, Newport Pagnell, Trowell and Rothersthorpe – have gained renown due to the adoption of their names for motorway service areas. In the case of Watford Gap, the service area and its location has come to assume an important symbolic position in the cultural (as well as physical and economic) geographies of the nation. As early as 1960, in an article entitled 'The Communications of Watford Gap, Northamptonshire' in the *Transactions and Papers* of the Institute of British Geographers, Hull geographer Jay Appleton explained why generations of engineers had routed 'at least half a dozen lines of communication', including canals, railways, roads and a motorway, through the Watford Gap, 'all of which have been, are, or will be of

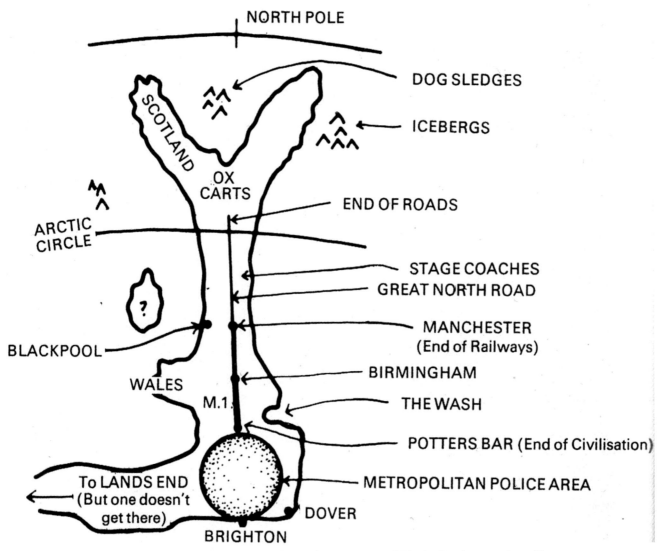

Figure 16.5 'How Londoners see the North – at least, according to the Doncaster and District Development Council'
Reproduced from Peter Gould and Rodney White's *Mental Maps* (1974) by permission of Rodney White and Penguin Books Ltd.

major importance in the economic geography of Britain'.[51] The physical geography of the region and specifically 'the Watford Gap' – comprising of two valleys cutting through the Northamptonshire Uplands – were identified as major forces in making this a prime location for so many major lines of communication passing between London and the South-East, and the Midlands and North-West. Appleton constructed Watford Gap as a key passage point in the economic, communications and physical geographies of the Midlands and Britain, and the M1/M6 corridor did quickly emerge as England's main North-South route. It was perhaps, inevitable that some commentators came to locate a national North/South divide on the M1 at Watford Gap, as it served as a resting stop just *beyond* South-East England, but *not quite of* the industrial landscapes of the Midlands and the North. This was a space beyond and between: a north/south divide that has been described in books as diverse as Rob Shields' *Places on the margin*, and Charles Jennings popular travel book *Up North: Travels beyond the Watford Gap*.[52] Popular references to Watford Gap appear to have a quite complex and somewhat unclear genealogy, with references to being 'north of Watford Gap' rarely mentioning the service area, and commentators often referring to being 'north of Watford' (which is actually 50 miles south of Watford Gap). Watford was frequently referenced as its position on the northern edge of London suggested a certain snobbery, narrow-mindedness and ignorance among metropolitan cultural commentators who refused to acknowledge England and Britain's cultural life outside of (and particularly north of) London. One interesting example of this is a map produced by the Doncaster and District Development Council that was reproduced in Gould and White's 1974 book *Mental Maps* (Figure 16.5). The Council produced 'Ye newe map of Britain' to illustrate 'how Londoners see the North'. On this map, the M1 is portrayed as the only transport link running up the centre of England: from London, through 'Potters Bar (End of Civilization)' (which is actually located near the A1, 8 miles east of the M1), to Birmingham.[53]

CONCLUSIONS

In this chapter I have used the metaphors of 'enfolding' and 'gathering' to approach the M1 as a topological and relational space. The construction of the M1 led to a reconfiguration of local, regional and national geographies, as particular places were rethought in relation to the location of the motorway and its junctions. The motorway itself was sited or 'placed'

in the landscape in different kinds of ways: as a structure to be seen; as a site to look out from, into the landscape and at particular landmarks; as a corridor to be developed; as a key north-south axis; and as a route on which are sited symbolic points in the geographies of Britain. Drawing upon Donald Janelle's writings on 'time-space convergence', David Harvey's writings on 'time-space compression', writings on 'tunnelling' and 'corridoring' effects, as well as broader work on mobilities by John Urry and others, we could frame some of these new relations and geographies in terms of new conceptions of distance, proximity, and connection, as a variety of cultural texts and the very physical infrastructure brought the motorway and its sites/sights 'close' to the lives of a variety of consumers: from rural residents and property developers, to pop enthusiasts, radio listeners and young children with toys and books.[54] But, as with many new forms of communication and transport, not everyone was brought 'close' to the motorway, and neither was this 'closeness' qualitatively uniform or universally desired. As traffic levels increased, many local residents resented their proximity to the noise, vibrations and pollution of the motorway. What's more, proximity to the motorway did not equate to good motorway access (i.e. to proximity to its junctions) or with positive 'compressive' effects on distance. As Doreen Massey famously remarked of David Harvey's writings, the processes and new technologies associated with 'time-space compression' do not have a homogeneous impact on all sections of society.[55] What emerges is a complex politics of mobility, depending upon whether one lives close to the motorway, has access to a private motor car, can afford to drive long distances, and a great many other social and cultural factors.[56] The motorway, here, unravels or unfolds through a variety of movements, media and material objects into people's everyday lives and the popular imagination in different ways, creating new geographies, new topologies.

ENDNOTES

1 Gilles Deleuze and Félix Guattari, *A Thousand Plateaus*, trans. Brian Massumi (London: Athlone, 1988), p. 363.

2 Ibid., p. 380, emphasis in original.

3 This was demonstrated in papers delivered at the 'Roads' symposium at the Department of Anthropology at the University of Aberdeen in 2006. On the anthropology of car use, see Daniel Miller (ed.), *Car cultures* (Oxford: Berg, 2001). The transience of some roads is also demonstrated in the popular US documentary-style television series *Ice Road Truckers*, which follows lorry drivers who transport goods to remote mines in Northern Canada along the seasonal ice roads over frozen lakes.

4 On the geography and history of England's M1 motorway see: Peter Merriman, *Driving Spaces: A Cultural-Historical Geography of England's M1 Motorway* (Oxford: Blackwell, 2007).

5 Limited access motorways possess both an unrelenting linearity and they generate distinctive 'tunneling effects' between junctions, creating an uneven geography of access. On the 'tunnel effects' created by motorways, see Paul Andreu, 'Tunneling', in Cynthia Davidson (ed.), *Anyhow* (Cambridge, Massachusetts: MIT Press, 1998), pp. 58–63; Stephen Graham and Simon Marvin, *Splintering Urbanism* (London: Routledge, 2001), p. 201.

6 There are a few rare exceptions, including: Peter Bishop, 'Gathering the land: the Alice Springs to Darwin rail corridor', *Environment and Planning D: Society and Space* 20 (2002), pp. 295–317; Michael Hebbert, 'Transpennine: imaginative geographies of an interregional corridor', *Transactions of the Institute of British Geographers* 25 (2000), pp. 379–392; Mitchell Schwarzer, *Zoomscape: Architecture in Motion and Media* (New York: Princeton Architectural Press, 2004).

7 Marc Augé, *Non-Places: Introduction to an Anthropology of Supermodernity*, trans. John Howe (London: Verso, 1995); cf. Peter Merriman, 'Driving Places: Marc Augé, Non-Places and the Geographies of England's M1 Motorway', *Theory, Culture, and Society* 21, 4–5 (2004), pp. 145–167; Peter Merriman, 'Marc Augé on space, place and non-place', *Irish Journal of French Studies* 9 (2009), pp. 9–29; Peter Merriman, 'Marc Augé', in Phil Hubbard and Rob Kitchin (eds), *Key Thinkers on Space and Place (Second Edition)* (London: Sage, 2011), pp. 26–33.

8 Kevin Hetherington, 'In Place of Geometry: the Materiality of Place', in Kevin Hetherington and Rolland Munro (eds), *Ideas of Difference: Social Spaces and the Labour of Division* (Oxford: Blackwell, 1997), pp. 183–199, quotes p. 184 and 187.

9 Marcus Doel, 'A Hundred Thousand Lines of Flight: a Machinic Introduction to the Nomad Thought and Scrumpled Geography of Gilles Deleuze and Félix Guattari', *Environment and Planning D: Society and Space* 14 (1996), pp. 421–439 (p. 421).

10 Peter Merriman, '"Operation motorway": landscapes of construction on England's M1 motorway', *Journal of Historical Geography* 31 (2005), pp. 113–133; Merriman, *Driving spaces*.

11 Peter Merriman, '"A new look at the English landscape": landscape architecture, movement and the aesthetics of motorways in early post-war Britain', *Cultural Geographies* 13 (2006), pp. 78–105.

12 Augé, *Non-Places*; Andreu, 'Tunneling'.

13 Automobile Association Archive, Basingstoke, Motorways file, 'Guide to the motorway' by Automobile Association, 1959.

14 'Motorway: *The Motor* guide to Britain's first long distance express highway', *The Motor*, 116, 4 November 1959, insert between pp. 508–509.

15 BBC Written Archives Centre, Caversham Park, Schools Microfilm 153/154, 23 June 1959–7 December 1959, 'Current affairs. No. 8. The new motorway' by Wynford Vaughan-Thomas, Transmitted 11 November 1959.

16 Royal Automobile Club, *Royal Automobile Club Guide and Handbook 1960* (London, 1960), p. 100.

17 Bishop, 'Gathering the land'.

18 John Postgate and Mary Postgate, *A Stomach for Dissent: the Life of Raymond Postgate* (Keele: Keele University Press, 1994), p. 269.

19 Raymond Postgate, *The Good Food Guide 1961–1962* (London: Cassel, 1961), p. 42.

20 Raymond Postgate, *The Good Food Guide 1965–1966* (London: Consumers' Association and Cassel, 1965), p. 187.

21 'Duke doesn't mind a land "grab"', *Luton News*, 27 June 1958.

22 Margaret Baker, *Discovering M1* (Tring: Shire Publications, 1968).

23 Ibid., p. 1.

24 Ibid., p. 3.

25 Augé, *Non-Places*.

26 Tony Brooks, 'The hazards of M1', *The Observer*, 8 November 1959, p. 5.

27 Cf. Andreu, 'Tunneling', p. 59.

28 Suzanne Greaves, 'Motorway nights with the stars', *The Times*, 14 August 1985, p. 8.

29 Eric Hartwell, 'Provision of catering facilities', *The Guardian*, 2 November 1959, p. 13.

30 See, for example, the collection of such postcards in Martin Parr, *Boring Postcards* (London: Phaidon, 1999).

31 'Trips to see motorway', *The Times*, 6 November 1959, p. 6; 'Minister "appalled" by new motorway driving', *The Times*, 3 November 1959, p. 8; '5,000 cars an hour on motorway', *The Times*, 9 November 1959, p. 10.

32 'M1' was released as a 7 inch single on Oriole Records, catalogue number 1573. It is a rather unusual, almost comic track. After listening to it I learned that Ted Taylor had gone on to act as musical adviser to the Benny Hill Show. I was not surprised.

33 D. Burke, 'The Ted Taylor Four story: an interview with Ted Taylor and Bob Rogers', *Pipeline: Instrumental Review*, 12 January 1992, pp. 1–10 (p. 6).

34 Robert Martin, *The Mystery of the Motorway* (London: Thomas Nelson and Sons, 1961).

35 Bruce Carter, *The Motorway Chase* (London: Hamish Hamilton, 1961).

36 Ibid., p. 19.

37 Ibid., p. 55.

38 L. Walker, D. Butland and R.W. Connell, 'Boys on the road: masculinities, car culture, and road safety education', *The Journal of Men's Studies* 8, 2 (2000), pp. 153–169 (p. 163).

39 Brian Dunning, 'The impact of the motorways', *Country Life*, 20 October 1966, pp. 978–979 (p. 978).

40 Ibid., p. 978.

41 Ibid., p. 979.

42 'Oakley give up land to avoid motorway', *The Times*, 6 August 1959, p. 15.

43 Dunning, 'The impact of the motorways'.

44 Malcolm MacEwen, 'Motropolis: a study of the traffic problem. Can we get out of the jam?', *The Architects' Journal* 130 (1959), pp. 254–275 (p. 270).

45 Ministry of Housing and Local Government, *The South East Study 1961–1981* (London, 1964).

46 Milton Keynes Development Corporation, *The Plan for Milton Keynes (Volumes One and Two)* (Wavendon, 1970).

47 'Survey: M1 corridor', *Estates Times*, 3 December 1993, pp. 10–34; Paul Sherman, 'Counties on the fast track', *Chartered Surveyor Weekly*, 23 May 1991, pp. 36–37.

48 Brian Turton, 'The road that started at Preston', *The Geographical Magazine*, 50, 7 (1978), pp. 452–456 (p. 453).

49 Daventry International Rail Freight Terminal 2006: DIRFT Logistics 2. Location. Website (accessed 3 July 2006), http://www.dirft.com/location.asp. The 'golden triangle' is formed in the space between motorways M1, M6 and M69, although it is also used to refer to surrounding areas.

50 Richard Osborne, 'The London-Yorkshire motorway: its route through the East Midlands', *The East Midland Geographer*, 13 (1960), pp. 34–38.

51 Jay H. Appleton, 'Communications in Watford Gap, Northamptonshire', *The Institute of British Geographers Publication No. 28: Transactions and Papers 1960* (1960), pp. 215–224 (p. 215).

52 Rob Shields, *Places on the Margin* (London: Routledge, 1991); Charles Jennings, *Up North: Travels Beyond the Watford Gap* (London: Little Brown, 1995).

53 Peter Gould and Rodney White, *Mental Maps* (Harmondsworth: Penguin, 1974), p. 40.

54 Donald Janelle, 'Central place development in a time-space framework', *The Professional Geographer* 20 (1968), pp. 5-10; Donald Janelle, 'Spatial reorganization: a model and concept', *Annals of the Association of American Geographers* 59 (1969), pp. 348–364; David Harvey, *The Condition of Postmodernity* (Oxford: Basil Blackwell, 1990); John Urry, 'Mobility and proximity', *Sociology* 36 (2002), pp. 255–274; Graham and Marvin, *Splintering Urbanism*.

55 Doreen Massey, 'Power-geometry and a progressive sense of place', in Jon Bird, Barry Curtis, Tim Putnam, George Robertson and Lisa Tickner (eds), *Mapping the Futures* (London: Routledge, 1993), pp. 59–69.

56 On the politics of mobility, see Tim Cresswell, *On the Move* (London: Routledge, 2006) and Tim Cresswell, 'Towards a politics of mobility' in this volume.

Coda

17   Landscapes on the Road

Gernot Böhme

Translated by Peter Cripps

Figure 17.1 Edvard Munch, Rue Rivoli *(1891), oil on canvas (photo by David Mathews)*

The film Lisbon Story by Wim Wenders (1994) begins with a fascinating prelude that in itself makes a substantial contribution to the film's theme, namely sound and image. The film's protagonist, a sound engineer, is driving by car from Berlin to Lisbon. He drives right across Europe – through Germany, France, Spain, Portugal, enjoying the new freedom from borders; 'This is my country, my homeland'. What unfolds in the process is Europe's diversity, as it reveals itself in front of his windscreen in terms of the various landscape types and city characters, and from the radio and the diverse linguistic sounds and songs that are broadcast. Yet this does not fully describe what fascinates us. For the fascination resides in particular in the gaze in motion – or to put it another way, in the view of the landscape that unfolds in motion. By now we are all well acquainted with such seeing in motion, even if we can only rarely afford to enjoy it for its own sake. In the film as well this is only possible because we are not doing the driving, in other words, we are 'relieved from action' and can thus allow the scenery simply to glide by. By now there are many films that make use of this travelling gaze, this seeing in motion. Indeed, there used to be, or still is, a film genre devoted to it: the road movie. Perhaps the best known example is Easy Rider, a film that uses the unimpeded gaze of the motorbike rider and, in this case, a deliberately leisurely style of riding, to celebrate the relation between freedom and landscape that marked the early years of the European landscape aesthetic. On the Copenhagen underground, one can study this moving gaze almost experimentally. Because the system is fully automatic and has no drivers, there is nothing to obscure the passenger's view into the tunnel from the front or the back of the train. Lines, contours and lights emerge from the darkness and float past in the corner of one's eye as close-up images. Or conversely, the picture shrinks, dwindles, is compressed, until finally it vanishes into the vortex of the black hole. But this is just a first step in tracking down the experience of moving landscape.

LANDSCAPE AS PICTURE

For the way we generally conceive of landscape is a picture, a static image.[1] We might go so far as to claim that for the average person the viewing of nature as landscape is learnt by means of landscape painting, or, today, photography. In the history of art it is claimed that the perception of a nature ensemble as landscape presupposes a landscape gaze, or, an eye for the landscape, to use the words of the early nineteenth-century aesthetic philosopher Wilhelm Heinrich Riehl:

> A landscape, as it presents itself to our gaze outdoors, is not beautiful per se, it merely has, as it were, the potential to be elevated to the beautiful and reshaped in the eye of the beholder. It is an artwork only insofar as nature provides the amorphous substance for such, while each viewer artistically moulds and animates that substance with the play of his eye.[2]

Of course, this conception is inspired by Immanuel Kant, according to whom the object of our experience is constituted by the forms of perception and categories of the subject. It is the subject who introduces unity into the manifold of the given, and, as the Critique of Judgement informs us, beauty is not a predicate of the thing but rather a quality of its effect on the subject. Philosophers and art historians have grounded this view on social historic considerations. Thus the art historian Matthias Eberle has shown that during the renaissance landscape was discovered by town dwellers, who,

already oppressed by the rules of civilized existence, sought their freedom outdoors, i.e. beyond the walls, and found it in nature qua landscape.[3] Examples include, first and foremost, those paintings in which the rural environment is depicted as the site of agricultural life, of nature and the seasons. Thus, moreover, the philosopher Joachim Ritter has claimed that the image of nature as landscape has a function that compensates for the analytical approach to nature, as this came to assert itself in the modern natural sciences.[4] In the landscape picture, nature could once again be perceived as a whole, as well-ordered and beautiful, in other words, as a cosmos. Examples include, first and foremost, early landscape pictures, which generally represented all that belongs to nature: plains and mountains, bodies of water from brooks to the sea, villages, towns, woods and fields. What one has to counterpose to this subjective and constructed conceptualization of landscape is the awareness that nature in its own right manifests characteristic landscape formations, and is thus not just an ensemble of uncoordinated elements. From communities of plants to rock formations and ecosystems, external nature possesses its own coherence. Alexander von Humboldt recognized this early on. For his part, he sought to integrate landscape painting into science as a means of capturing the overall impression of a landscape. One of the people he engaged for this work was the early romantic painter, Josef Anton Koch. Conversely, scientists such as Ernst Haeckel and Wilhelm Oswald themselves painted landscapes in the course of their researches. Certainly, Humboldt was concerned with the overall impression of a landscape as experienced by the human individual, nature as it appears to us, in other words, anthropo-geography. Thus what counted even for these scientifically oriented and objectively schooled authors and painters was landscape in the sense of a picture, a prospect.

This notion of landscape is clearest where it is reflexively indicated within the painting as such, as it is in the works of Caspar David Friedrich. In many of his pictures, Friedrich includes a centrally placed observer – sometimes several of them – his famous back-view figures (Rückenfiguren). It

Figure 17.2 Sunset with Rückenfiguren, Kamakura, Japan (photo by Gernot Böhme)

has been pointed out that these figures are more than just staffage, in other words, they are more than mere foreground accessories or attempts to enliven a picture with elements of everyday life.[5] No, these back-view figures are manifest observers, people generally shown facing the landscape in either contemplative or expectant mood. As examples of the latter one could mention the moonrise pictures, while examples of the former are Der Wanderer über dem Nebelmeer [The Wanderer Above the Sea of Fog] from around 1817 and, above all, Der Mönch am Meer [The Monk by the Sea] from 1810. Der Mönch am Meer was the first painting that Friedrich imbued with this quasi reflexive dimension through the introduction of an observer. The recognition of this feature, and hence also of a painterly innovation, is attributed to Heinrich von Kleist, or alternatively, to the romantic triumvirate of Kleist, Clemens von Brentano and Achim von Arnim.[6] In his short newspaper article 'Empfindungen vor Friedrichs Seelandschaft'[7] ['Impressions on Viewing Friedrich's Seascape'] we find the phrase, 'and thus I myself became the Capuchin'.[8] This pins down the back-view figure as a subject with whom the individual observer identifies.

Friedrich's back-view figures usually perform a strongly organizing function. They provide a kind of vanishing point for what is presented in the foreground of the picture. Thus they serve as identification figures for the viewer of the landscape picture. This is particularly so in Der Mönch am Meer, because here the tininess of the figure renders his loneliness within the scene palpable for the viewer. And a scene it is, even so. For conversely, the back-view figure creates a distance for the person looking at the picture; what one is looking at here is not nature, but a view of nature. Nature as image, nature as drama. The Capuchin, the monk by the sea, is a forerunner, as it were, of the modern individual. Although the latter might stand in the midst of nature, he keeps it at a distance by photographing it.

The influence of landscape painting on the aesthetic concept of landscape can be traced all the way down to our modern assessment of landscape, and the shaping of landscapes for ecological and touristic purposes. One early indication of this is that, in his aesthetics of nature, Immanuel Kant described the landscaped garden as a landscape painting rendered concrete.[9] In recent discussions, the most notable exponent of this approach is the landscape architect Werner Nohl, whose empirical investigations of landscape are based exclusively on evaluations of landscape photos, in other words, static images.[10] Similarly, landscape tourism, which began to develop in Europe in the eighteenth century, was from the outset concerned with vedute, with vistas and prospects. Thus even today car parks with adjacent viewing points are a frequent feature along roads of particular scenic interest. The landscape is a picture, which the tourist has to be enabled to register, so as to be able 'to carry it safely with him home'.[11]

However, given this traditional, painting-oriented landscape aesthetic, two questions arise.

a) Isn't our predominant mode of landscape perception one of movement? Hasn't our aesthetic enjoyment of landscape for many years now been one of travelling through it? Is landscape as perceived during travel really nothing more than a rapid sequence of images that we seek in vain to capture – or is there such a thing as a flow perception of landscape?

b) Is landscape, at least in the European region, still really authentic nature, the appeal of which, as Joachim Ritter and Matthias Eberle believe, consists in its contrast to the civilized world? On the one hand, isn't it our human impact that characterizes the landscape, in the past in the form of traditional agriculture, today in terms of technical infrastructure? And on the other, don't devastated landscapes and the new wastelands resulting from industrial production possess a certain appeal?

LANDSCAPE IN MOTION

We must ask how technology conditions our perceptions of landscape. More precisely, we must ask how a viewer in motion perceives the landscape, in particular the landscape perception of someone travelling by train or by car. There are two seminal titles on this theme: Wolfgang Schivelbusch's history of rail travel, The Railway Journey (in German 1977) and Marc Desportes' Paysages en movement [Landscapes in Motion] (2005).[12] The substance of these books as relevant to our own discussion deserves a brief summary.

The authors maintain that such a thing as landscape was first discovered on the one hand by and on the other for the traveller as a result of technically improved means of transport. But the first claim is not entirely true, because even early tourists – most of them English, although the wanderings of the young Goethe also deserve mention – who generally travelled on foot, had an eye for the landscape.[13] The second claim, however, is more convincing; it was improvements in road conditions and coach construction in the eighteenth century that made it possible for travellers to let their gaze wander about outside without fear and constant jolting.

More important is the claim that landscape perception changed as means of transport developed. After describing the irritations the traditional traveller had to endure, Schivelbusch notes a transformation of experience. 'The velocity and linearity with which the train traversed the landscape did not destroy it – not at all; only under such conditions was it possible to fully appreciate the landscape'.[14] He sums up one particular aspect of the landscape gaze that rail travel facilitated under the term 'panoramic seeing'. This is characterized first and foremost by the diminished significance of the foreground, which has to be overlooked due to the speed. This loss entails at the same time the loss of a physical relationship to one's immediate surroundings. Schivelbusch even speaks of alienation; one no longer has any sensory contact with the surrounding landscape. The more comfortable the travel, the more it becomes a purely visual experience. This tends to create still greater unity among the areas in the distance, in other words, it unifies the landscape, which reveals itself as 'a sequence of images or scenes', insofar as it introduces the traveller to 'ever-changing perspectives'.[15] The disappearance of the foreground also presents the experienced rail traveller with new visual experiences, such as that of the blurred or smeared object. Certainly, it is not something that everyone wants to indulge in, yet the foreground zipping by presents even the traveller who is reluctant to look with an unusual experience, namely that of ephemeral impressions, which can be characterized as a visual experience similar to the snapshot, a mode of photography that first arose in the twentieth century, whereby the one who is speeding past captures the momentary gesture of a person, an animal, or a scene.[16]

So much for rail travel. During the early years of car travel, according to Desportes, the physical-practical relation to the landscape was to some extent reaffirmed. Driving becomes a form of corporeal experience, with the conquest of the landscape by means of the car constituting an individual undertaking. This changed with the advent of purpose-built roads, especially with the construction of motorways. On the latter, the width of the carriageway is in itself an obstacle to the landscape gaze; what dominates is the view of the road, which, moreover, takes its orientation less and less from objects, such as buildings and landmarks, as it

did in the past, and increasingly from signs. Admittedly, there remains a diffuse perception of impressions to either side, but the principal focus of concentration is now straight ahead. 'Un monde frontal' is the heading Desportes gives his chapter covering the years 1920–40. Thus the highway in its turn brings about an alienation of the individual from the landscape. Even so, Desportes notes a positive transformation of perception, insofar as and because one is not so rigorously focused on what is functional when on a motorway:

In viewing the features of the highway and the sights it presents to us as a stream framed by the windscreen, the spectacle appears as a rhythmical sequence. The motorist sees these elements coming towards him, then passes them by, while discovering others. It is a spectacle that can afford genuine pleasure; the changing perceptions of things, their relative movements, the game of coming into view and vanishing are all sources of delight.[17]

We should note that, under the heading 'Le rythme retrouvé', Desportes interprets the positive variant of landscape perception associated with car travel as a kind of filmic seeing. This explains why he precedes this section with an interlude on the theme of cinematography.[18]

RAPID CUTS AND FLOW PERCEPTION

Both of the authors just mentioned take on greater significance with the major contributions that they make to our understanding of the changes in human perception through social and technological history. With regard to landscape perception in technological civilization, it could be, however, that they are not radical enough. The questions they ask still seem to revolve around how and when we perceive landscape in the traditional sense of the word. On closer consideration, they are still more conservative, insofar as they take landscape to be something out there. This conception is implied by their claim that rail travellers or car drivers are alienated from the landscape. It is true that such alienation from the physical environment does occur, but it is precisely for this reason that the landscape becomes a picture. This was the lesson that we could draw from Friedrich's paintings, which constitute reflexive landscape paintings, insofar as his back-view figures explicitly present the nature 'out there' as something seen. The point is that Friedrich's back-view figures are not integrated into the landscape – as is, for example, the shepherd in his painting Der einsame Baum [The Lonely Tree]. With respect to the landscape, they retain a distance that corresponds to that of Kant's aesthetic subject facing the sublime, which, objectively, is overwhelming and all-powerful nature.

The panoramic gaze of the average rail traveller, like the cinematic gaze of the average car driver, is an instantiation of a view of nature, or more specifically of nature as landscape, that has always informed the practice of the landscape painter. With the difference that here the perceived landscape pictures are no longer static images, but amount rather to a sequentially scanned panorama in the former case and to an experience of things appearing and disappearing in the other. In other words, these are cases of landscape in motion – and it is with good reason that these are the words Desportes uses for the title of his book. At the same time it was the smoothness of the motion and the speed of these technically advanced means of transport that made this gaze possible. Because, obviously enough, foot travellers, horse riders and early coach drivers were also in motion. But for them, the sequence of views that they registered along the way did not arrange themselves into a unity.

Landscape in motion is something different from landscape as picture. Our understanding of how celluloid film works tends to confuse us here; we are aware that moving pictures consist of a rapid sequence of static images. Correspondingly, the landscape that emerges into view from the distance is seen by analogy to the zoom, in other words, as a static image that we draw in towards us.

This approach cannot be applied, however, to the impression one has of the foreground rushing past when looking from a train or through the side window of a car. For here the impression is such that one cannot distinguish individual details, which, as noted above, are reduced to a blur. We can compare this form of seeing to what is called flow perception, a sensory mode ascribed to insects. This is a type of perception that provides orientation not so much in terms of positioning within a constellation of objects but rather by means of a dynamic space.[19] Of course, one has to assume that, as slow-moving creatures, humans prefer to perceive pictures and tend to avoid flow perception as something less pleasant. Yet the question we are dealing with is whether technological lifestyles acquaint us with entirely different modes of perception. And it could be that this is precisely what car driving does for us. It is interesting to note, incidentally, that painters have attempted to present the impressions of flow perception in pictorial form – for example, Edvard Munch in his *Rue de rivoli* (1891) or William Turner in his *Snow Storm – Steam-Boat off a Harbour's Mouth* (1842).

There is another mode of perception, which one could call fractal. Under the heading 'Un espace fragilisé', Desportes draws attention to the fragmentation that can affect the motorist's perception – albeit with the result that for the motorist the rapid changes in scenery prevent a landscape picture from ever taking shape. Here we have to note, firstly, that the liberated gaze, which in Desportes' view gives rise to a filmic perception of landscape, is in fact only really possible for a passenger. And secondly, that the driver's gaze tends to be far more fragmented than Desportes acknowledges. For we have to consider the numerous glances at the rear-view and wing mirrors that are absolutely indispensable to someone driving on the motorway. The type of perception that arises here can only be compared to that of rapid cutting in a film, a style familiar from video clips. What is thereby produced is certainly not a picture, nor even a conscious impression of a whole, which, in the case of flow perception, it might still be possible to capture in suggestive pictorial form. Nevertheless, precisely because a video clip does not really amount to a film, it is able to convey an overall atmosphere, even if this is something heavily dependent on the interplay of sound and sight. And similarly, the constantly shifting gaze of the motorist leaves him with a certain sense of the landscape, or rather, of an atmosphere that is the product of the season, the wind and the weather, the landscape surroundings, and traffic conditions. Perhaps one could say that, just as the train passenger becomes aware of the distant landscape due to the failure of the foreground to yield a picture, the motorist becomes aware of the background atmosphere that surrounds him due the failure of the rapid cuts to produce a picture.

Undoubtedly, in both cases he will be grateful to be relieved of these impressions. It will be a joy to pause for a while at a conveniently situated picnic area and, with a sigh of relief, to let his gaze dwell meditatively on the scenery of a landscape.

ENDNOTES

1   Norbert Bolz and Ulrich Rüffer (eds), *Das große stille Bild* (Munich: Fink, 1996).

2   Wilhelm Heinrich Riehl, *Kulturstudien aus drei Jahrhunderten* (Stuttgart: Cotta, 1859), p. 67. See also the chapter: 'Das landschaftliche Auge', pp. 57–79.

3   Matthias Eberle, *Individuum und Landschaft: zur Entstehung und Entwicklung der Landschaftsmalerei* (Gießen: Anabas-Verlag), 1980. Characteristically, Eberle uses Friedrich's *Der Wanderer über dem Nebelmeer* (1817) as the book's cover image.

4   Joachim Ritter, *Landschaft. Zur Funktion des Ästhetischen in der modernen Gesellschaft* (Münster: Aschendorff, 1963).

5   Ewelina Rzucidlo, *Caspar David Friedrich und Wahrnehmung. Von der Rückenfigur zum Landschaftsbild* (Münster: LIT, 1998).

6   Werner Hofmann, *Caspar David Friedrich. Naturwirklichkeit und Kunstwahrheit* (Munich: Beck, 2000), ch. 2, 'Der Mönch am Meer und Die Abtei im Eichwald', pp. 53–82.

7   Friedrich did not usually give his paintings titles. What Kleist refers to here as Seelandschaft [Seascape] is the painting that has come to be known as Der Mönch am Meer [The Monk by the Sea].

8   Heinrich von Kleist, *Sämtliche Werke und Briefe* (Darmstadt: WBG, 1962), vol. II, p. 327. My formulation here is as cautious as possible, because, as a whole, Kleist's text is hard to understand. The sentence quoted might also mean that Kleist as viewer of the picture – rather than of the sea – becomes the Capuchin.

9   Immanuel Kant, Critique of Judgement. §51, A208f.

10 Werner Nohl and Klaus Neumann, *Landschaftsbildbewertung im Alpenpark Berchtesgaden – Umweltspsychologische Untersuchungen zur Landschaftsästhetik. MAB-Mitteilungen* (Schriften-Reihe Deutsches Nationalkomitee für das UNESCO-Programm 'Der Mensch und die Biosphäre'), no. 23. Bonn: 2nd ed., 1986. By the same authors, 'Zur Rolle des Nicht-Sinnlichen in der landschaftsästhetischen Erfahrung', in *Natur und Landschaft* 65 (1990), no. 7/8, pp. 366–370. See also Adrian von Buttlar, 'Gedanken zur Bildproblematik und zum Realitätscharakter des Landschaftsgartens', in *Die Gartenkunst*, no. 1, 1990, pp. 7–19.

11 Johan Wolfgang von Goethe, *Faust I*, verse 1968.

12 Wolfgang Schivelbusch (Berkeley, California: University of California Press, 1986). Marc Desportes, *Paysages en mouvement. Transports et perception de l'espace, XVIIIe– XXe siècle* (Paris: Gallimard, 2005).

13 Goethe sat and sketched at the Gotthard Pass. See *Dichtung und Wahrheit*, Part 4, beginning of Book 19.

14 Schivebusch, *The Railway Journey*, p. 59.

15 Ibid., p. 61. In fact, this landscape gaze is closer to the experience of film than to that of the panorama, which represents a distant and sweeping vista, yet one that is

static. Although Desportes notes this point, he still adopts the idea of the panoramic landscape. Desportes, *Paysages en mouvement*, p. 147.

16 This trend first became popular with the availability of Kodak's Browny box camera around 1900.

17 Desportes, *Paysages en mouvement*, p. 336. 'C'est en voyant défiler, cadres par un pare-brise, les composantes de l'autoroute et les sites qu'elle donne à voir que ce spectacle peut être reçu comme une séquence rythmée. L'automobiliste voit venir à lui des éléments, les dépasse, en découvre d'autres. Un véritable plaisir est lie à ce spectacle: la vue changeante des choses, leurs mouvements relatifs, leurs jeux d'apparition et de disparition sont autant de sources de joie'.

18 Ibid., pp. 243–254.

19 The easiest example is the identification of the direction of flight by means of the point in the field of view at which the focus opens. (I thank my daughter, Rebecca, for the information on optic flow.)

Index

Bold page numbers indicate figures, *italic* numbers indicate tables.